Sandra van Soelen

From violent conflict to peaceful cooperation

AF153991

Sandra van Soelen

From violent conflict to peaceful cooperation

The nature of irrigation management situations in Northern Ethiopia, the role of institutions and lessons learned

LAP LAMBERT Academic Publishing

Impressum / Imprint
Bibliografische Information der Deutschen Nationalbibliothek: Die Deutsche Nationalbibliothek verzeichnet diese Publikation in der Deutschen Nationalbibliografie; detaillierte bibliografische Daten sind im Internet über http://dnb.d-nb.de abrufbar.
Alle in diesem Buch genannten Marken und Produktnamen unterliegen warenzeichen-, marken- oder patentrechtlichem Schutz bzw. sind Warenzeichen oder eingetragene Warenzeichen der jeweiligen Inhaber. Die Wiedergabe von Marken, Produktnamen, Gebrauchsnamen, Handelsnamen, Warenbezeichnungen u.s.w. in diesem Werk berechtigt auch ohne besondere Kennzeichnung nicht zu der Annahme, dass solche Namen im Sinne der Warenzeichen- und Markenschutzgesetzgebung als frei zu betrachten wären und daher von jedermann benutzt werden dürften.

Bibliographic information published by the Deutsche Nationalbibliothek: The Deutsche Nationalbibliothek lists this publication in the Deutsche Nationalbibliografie; detailed bibliographic data are available in the Internet at http://dnb.d-nb.de.
Any brand names and product names mentioned in this book are subject to trademark, brand or patent protection and are trademarks or registered trademarks of their respective holders. The use of brand names, product names, common names, trade names, product descriptions etc. even without a particular marking in this work is in no way to be construed to mean that such names may be regarded as unrestricted in respect of trademark and brand protection legislation and could thus be used by anyone.

Coverbild / Cover image: www.ingimage.com

Verlag / Publisher:
LAP LAMBERT Academic Publishing
ist ein Imprint der / is a trademark of
OmniScriptum GmbH & Co. KG
Heinrich-Böcking-Str. 6-8, 66121 Saarbrücken, Deutschland / Germany
Email: info@lap-publishing.com

Herstellung: siehe letzte Seite /
Printed at: see last page
ISBN: 978-3-659-68022-9

Abstract

Water is crucial to human life and the earth as a whole; there is no life without water. The majority of the rural population in Ethiopia is depending on agriculture for their livelihood and most agriculture is still based on the availability of rain. This limits the possibilities of the rural population to have a consistent income all year long; their agricultural practices are more focused on subsistence farming. Irrigation developments are growing rapidly in Ethiopia and there is a water resource potential to increase the irrigated agriculture substantively over the coming years. Managing these water systems is however not always easy. Water management equals managing conflicting interests and farmers can get competitive to secure their own access to water for irrigation. However, there is not much knowledge available about the nature of these local management situations. Therefore, a research has been conducted to look at the nature of local management situations, the role of various institutions and stakeholders, and the way in which problems are dealt with (conflict management). During three months of fieldwork in Ethiopia, various locations have been visited and different cases have been investigated. These cases are used in this thesis to describe the main trends in Ethiopia and to give recommendations on how to move forward. During the fieldwork period, semi-structured interviews have been conducted with various stakeholders in the cases itself, as well as government officials on regional and federal level. The aim was to know the contextual factors of the conflicts, the stakeholders involved and the dynamics around the situation. In total, seven cases have been selected which are located in three different regional states in Ethiopia.

From the research, it appeared that the developments in irrigation are in general positive for the livelihoods of the rural poor, but there is a lack of attention for irrigation management and there are often local conflicts or situations of void present (situations where there is no conflict or cooperation, but the water resource can be deteriorated or developed blindly). These situations can threaten the sustainability of the local systems. Most of the conflicts are on local scale, around one system, between farmers from two communities or between farmers and the local government. Management situations are dynamic situations which can change from nature over time, due to climate variability (there are more conflicts in the end of the dry season than during the rainy season) or changes in the social-political situation. The conflict situations are not caused by one factor (such as scarcity in the dry season), but often by a combination of different factors, whereby the functioning of institutions is often involved. This is related to the functioning of local water committees responsible for the day to day management of irrigation systems, as well as the local government supporting the water committees and often responsible for mediating in local conflicts. Most of the conflicts are dealt with on local level, without much supervision from guidelines or procedures on how to deal with conflicts. As a result, the approaches used by the different local government authorities differ, creating different results. Since these local conflict situations are present, attention for conflict management and prevention are needed in the management approach of local irrigation systems: attention during trainings of local water committees and local government authorities is needed to make them more aware about these situations and how to deal with them.

Key words: irrigation management, water-related conflicts and cooperation, institutions, conflict resolution and prevention.

All pictures used in this thesis are made during the fieldwork period in Ethiopia by the author.

Acknowledgement

The experience in Ethiopia learned me a lot about conducting fieldwork and living in a foreign country. Some research situations were a bit strange and unexpected. For example I conducted an interview along the main road from Addis Ababa to Mekelle on a Sunday morning and every time when a truck of bus passed, the interview had to stop for a second. When I had a car for ten days to conduct fieldwork, we also drove through dry rivers beds to visit villages and irrigation schemes. As it is known for Africa, I also conducted interviews underneath trees and in the middle of agricultural fields sitting on recently ploughed land. I visited different governmental offices and saw a big variation in furniture and assets between the different areas and levels of government. The people in Ethiopia were in general very kind and willing to help me during my research. This made my time in Ethiopia unforgettable and a good input for this master thesis.

During my research, I had a lot of help from people in the Netherlands, but especially in Ethiopia. I performed this research for MetaMeta and I would like to thank the whole organisation for realising the opportunity to do fieldwork in Ethiopia and assisting in my research, especially Frank van Steenbergen, Asefa Kumsa, Taye Alemayehu who supervised me during my research.

During my three weeks in Mekelle I had a lot of help from professors from the Mekelle University: especially Dr. Kifle Woldegeray and Dr. Dessie Nedaw Habtemariam, who went with me to the field. Also Dr. Eyasu and Dr. Seifu Kebede (Addis Ababa University) assisted me during the research.

In one part of my fieldwork I got assistance of the local NGO Water Action; I would especially like to thank Berhanu and Temesgen for their time and resources and making this research opportunity possible. The translators in Degem, Diriba Mengistu, Bekele Feyissa and Bekele Amante, also helped me for a part of my fieldwork. I would like to thank Misgan Adino and Tilaye Bitew Bezu for keeping in touch and helping me with all my questions. And I would like to thank all farmers, government officials and other respondents who answered all my questions during the fieldwork in Ethiopia.

Besides conducting my research, I had a wonderful time in Ethiopia. A special thanks to of the guesthouse I stayed in Addis Ababa and all my friends in Ethiopia for making my time in Ethiopia unforgettable!

At Utrecht University, I would like to thank Dr. Joris Schapendonk for supervision on my thesis for his good advice during the process. This helped me in improving the story I am telling in my thesis. Lastly, I would like to thank my family and friends in the Netherlands for supporting me and giving me advice during the process and my whole master programme.

Sandra van Soelen
sandravansoelen@gmail.com

Utrecht, August 2013.

Contents

List of figures, maps and tables

Abbreviations

ADSWE: Amhara Design and Supervision Works Enterprise
CIA: Central Intelligence Agency of the United States
CPR: Common Pool Resources
DA: Development Agent
DFID: Department for International Development, United Kingdom
ECWP: Ethiopian Country Water Partnership
FDRE: Federal Republic of Ethiopia
GDP: Gross Domestic Product
GTP: Growth and Transformation Programme
GW-MATE: GroundWater Management Advisory Team
GWP: Global Water Partnership
Ha: Hectares
IWRM: Integrated Water Resource Management
KGVDP: Kobo-Girana Valley Development Programme
MDG: Millennium Development Goals
MoFED: Ministry of Finance and Economic Development
MoWE: Ministry of Water and Energy
MoWR: Ministry of Water Resources
NGO: Non-Governmental Organisation
OECD: Organisation for Economic Co-operation and Development
REST: Relief Society of Tigray
UK: United Kingdom
UN: United Nations
UNDP: United Nations Development Programme
WASH: Water, Sanitation and Health
WFP: World Food Programme

1. Introduction

"The water problems of our world need not be only a cause of tension; they can also be a catalyst for cooperation. If we work together, a secure and sustainable future can be ours" (Kofi Annan, 2002).

Water is crucial to human life and the earth as a whole; there is no life without water. Water is necessary for nearly every sector of human activity. Unfortunately, water is a finite and vulnerable natural resource, unevenly distributed in space, time, quantity and with great variations in quality. In the past hundred years, the global population has tripled while demand for water has increased seven-fold. The current influence of climate change is also resulting in increasing prevalence of droughts, floods and other weather events (Malley et al., 2009, p.12). As a consequence, it is predicted that 50% of people in Africa by 2025 will face water stress and scarcity (Ibid, p.4). As a result, water has been high on the international political agenda for a long period of time.

Millions of the world's poor, particularly in rural areas who are living from subsistence agriculture, depend on water for their livelihood. The agricultural sector is highly dependent upon water; globally it is estimated that 10% of world water withdrawal is used for domestic uses, 20% for industrial uses and 70% for irrigated agriculture (OECD, 2005). Water scarcity, in combination with low economic development and shortfalls in political governance may lead to instability, threatening rural livelihoods (Ibid). The availability of water, now mostly provided by rainfall, determines the success of the local farmers. However, rainfall equals major variability in most parts of Africa making farmers vulnerable for variations and shocks. On the positive side, the development of irrigation and agricultural water management holds significant potential to improve productivity and reduce vulnerability to variable climate conditions (Awulachew, 2010). Different water sources can be used for irrigation developments. Groundwater developments have created miracles in accelerated agricultural production in rural India, China, North Africa and the Middle East (GW-MATE, 2011). However, large scale groundwater developments are not present in Sub-Sahara Africa.

In all societies, water is a powerfully unifying resource, but because of its centrality to human life and our ecosystem, its management is generally diffuse (Carius et al, 2004). Water management is highly complex and extremely political. There is also a growing competition for water resources, caused by factors such as a growing population, changing water use patterns, new types of water use and climate change (Ravnborg et al., 2012). These factors contribute to a higher risk of water-related conflicts on small geographical scale. There are also situations were cooperation is the result of pressure on the water resources. However, knowledge about these intra-community conflicts is limited and most of the conflicts remain undocumented (Ibid). Currently, most of the research focuses on trans-boundary water conflicts; there is less research done on the nature of local water conflict (Ibid/Funder et al., 2010). The particular strategies of the poorest community members in water conflict and cooperation remain relatively poorly understood (Funder et al., 2012). The knowledge about the local conflicts and cooperation can be input for the formulation of new policy, legal and institutional frameworks for water governance and management that is currently taking place in many developing countries (Ibid). To increase the knowledge of local water-related events, the focus of this thesis is on Ethiopia; one of the countries were rapid developments are taking place around irrigation and were these developments are leading to small scale conflicts which remain unknown to most of the politicians.

Ethiopia is a poor country in the Horn of Africa where around 80% of the population is depending on agriculture for their livelihood. The agricultural sector is characterised by small scale farmers that predominantly make use of traditional farming methods (Van Koppen et al., 2009). In the past, there have occurred some major droughts in various parts of the country, examples are the disaster periods of 1973-74 and 1984-85, which received major international attention (Sørenson and Bekele, 2009). Some parts of Ethiopia still experience periods of droughts and are many household depending of food assistance. This situation is negatively influencing the development of the country. According to study conducted by the African Union Commission and the UN World Food Programme (WPF), child under-nutrition was leading to a loss of $4,7 billion for the Ethiopian economy in 2009. Two out of five children are stunted in Ethiopia. This is leading to 20% of child deaths and has reduced the Ethiopian workforce by 8% (Schlein, 2013). The government of Ethiopia is trying to change this situation, by promoting water-centred development, which is seen as entry point for growth and improved livelihoods in Ethiopia (GW-MATE, 2011).

Ethiopia has abundant rainfall and water resources, but the current agricultural system does not fully benefit from developments in irrigation (Awulachew, 2010). Currently, surface water resources are predominantly used for irrigation systems, although groundwater is widely available in Ethiopia (GW-MATE, 2011). Next to areas where groundwater is intensively used (in most cases very recently), there are large potential area where it is not extensively used at all. This is not unlike other countries in Sub-Saharan Africa. The expectation is that this will change, especially in areas with very shallow groundwater resources (GW-MATE, 2011). However, there are local conflicts and problems present around water systems, which are threatening the sustainability of the water systems and are not contributing to development (Ibid; Butterworth et al., 2011). In discussion on the new Growth and Transformation Plan, the Government of Ethiopia singled out groundwater development as one of the major drivers for agricultural transformation (MoFED, 2010b). The overall intention is to reach 5 million hectares under irrigation at the end of the second Growth and Transformation Plan (GTP) in 2020. Groundwater is also basis for the water supply program. The Universal Access Plan has high ambitions – reaching 100% coverage for urban areas in 2015 and 98% for rural areas – all to be met from groundwater (MoFED, 2010b).

1.1 Research questions

This thesis describes the results from a master research done in cooperation with MetaMeta and specifically for the CoCoon project; groundwater in political context. The aim of this project is to explore the dynamics of cooperation and conflict over natural resources. This research is focusing on different management situations around irrigation systems in Northern Ethiopia. The problems and conflicts around local irrigation systems are hardly documented and barely known by regional and national government bodies, as well as situations of cooperation (Van Steenbergen, 2011/Ravnborg et al., 2012). Besides describing the cases, there is a focus on the role of institutions and the management of irrigation systems. There are a lot of systems being constructed, but the functionality over time is often an issue. Major problems which are leading to a lack in sustainability are poor sense of ownership, lack of transparency and accountability of water committees, lack of available spare parts and standardisation of schemes (Kebede, 2013). For this research, an extensive effort was made by contacting local resource persons to identify cases of conflict or cooperation related to irrigation systems. The aim of this research is to increase the knowledge of local water management situations by describing and analysing empirical management cases in Northern Ethiopia. The main research question is: *What is the nature of situations of conflict, cooperation and void around irrigation schemes in Ethiopia, who intervened and what lessons can be learned from these situations?* This thesis will focus on local management of irrigation schemes and the occurrence or absence of local conflicts around irrigation systems. The following three sub-questions were formulated to guide the research and will be central in the empirical chapters:

1. *What are the characteristics of local management situations and which factors contributed to these situations?* This sub-question will look at the characteristics of the management situations; what is the background of the situation, what are problems and issues at stake. Moreover, the question will look at the factors which have contributed to the different situations.

2. *What is the role of local government institutions in the local management situations?* The local government is playing an important role in most of the local management situations. The formal roles and responsibilities of government organisations will be described, as well as the role of local water committees, who are responsible for the day-to-day management.

3. *What kind of conflict management tools can be used to solve local conflicts?* There are different conflict management tools and preventive measures that can be used to solve local conflicts. These tools and measures will be compared by means of discussing approaches used in the empirical cases. This will provide insight to give (policy) recommendations for Ethiopia.

1.2 Readers guide

In this thesis there is first attention for the background, theory and design of the research in chapter 2, 3 and 4. In chapter 2 the regional background is described, with attention for the characteristics of Ethiopia, the water resource policies and the main developments in irrigation in the country. This will give context information and will illustrate the main developments in Ethiopia. The third chapter will focus on the theory used in this thesis and will end in a theoretical framework, which is the basis for the empirical analysis of the data collected in Ethiopia. In this chapter, the main concepts of water conflict and cooperation, institutions and conflict management are discussed. The fourth chapter will elaborate on the methodology used for this research, including a description of the data collection period and the limitations of the research. A conflict analysis model of the British Department for International Development (DFID) is used as basis for the research approach.

Chapter 5, 6 and 7 are the three empirical chapters, based on the data collection in Ethiopia and discussing the three sub-questions. For the thesis, seven cases of irrigation management are chosen to illustrate the diversity of situations and how different organisations play a role in these situations. In chapter 5, these cases will be introduced, with a focus on the characteristics of the cases and an analysis on the influences of factors influencing the situations identified in the theory chapter. Chapter 6 and 7 are using the same cases, but discuss more specific subjects related to the second and third sub questions. Chapter 6 will focus on the management of irrigation systems; the roles and responsibilities of different government organisations and their influence in the empirical cases. Three cases will be discussed in detail, whereas the other cases are input for the synthesis of the chapter. Chapter 7 will look at conflict resolution, which is part of management, but not used in everyday management situations. How this is implemented or not in Ethiopia and what the way forward is, discussed based on the experiences from three of the cases were a form of conflict resolution is used. The thesis will end will a concluding chapter, reflecting back in the main research question and giving recommendations for improvements in Ethiopia.

2. Regional background

The horn of Africa is known for its frequent droughts and erratic rainfall. The major Ethiopian disaster periods of 1973-1974 and 1984-1985 were induced by complete crop failure and major death of livestock (Sørenson and Bekele, 2009: 23).

Ethiopia is a landlocked country in Eastern Africa with a population around 94 million people (CIA, 2013). Neighbouring countries are Eritrea, Djibouti, Somalia, Kenya, Sudan and South Sudan. There are different ethnic groups living in Ethiopia, the two biggest are Oromo 34.5% and Amara 26.9% (CIA, 2013). Ethiopia has a Human Development Index of 0.383, which represents the 174th place in the ranking (UNDP, 2012). Life expectancy at birth is 56.56 years (CIA, 2013). As in other countries in Africa, the majority of the people living in the country are depending on agriculture. The agricultural activities account for 41% of the GDP, 60% of export and 80% of the working force (CIA, 2013/Van Koppen et al., 2009). Droughts and poor cultivation practices heavily influence the agricultural sector, although the situation is changing. In the period 2009/10, 29.2% of the population was living underneath the poverty level of $1.25 a day (MoFED, 2010a, p.5). According to the World Food Programme, 3.7 million people were depending on food aid in 2012[1] (WFP, 2013).

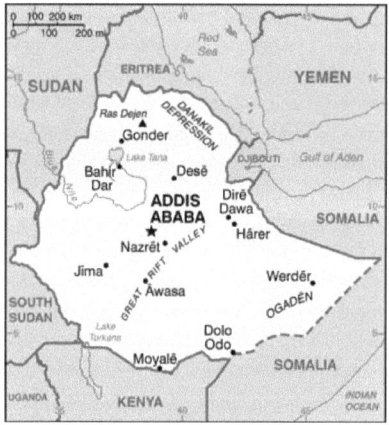

Figure 2.1: Map of Ethiopia (CIA, 2013)

Ethiopia has a rich history and a unique position in the history of Africa of remaining independent during the time of European colonisation in Africa; there was only a short invasions of the Italians between 1936 and 1941 (Henze, 2000). In the past decades, the politics and leadership of Ethiopia have changed from time to time. One of the most famous emperors was Emperor Haile Selaissie, who had the power in Ethiopia from 1930 until 1974 with an interruption of five years due to the Italian invasion. He was the last emperor and his successful leadership ended due to a period of food shortages and border wars. A communist military junta led by Mengistu disposed Haile Selaissie

---

[1] The Ethiopian calendar is 7/8 years behind the Gregorian calendar we use in the Netherlands. In this thesis, the years have been converted into the Gregorian calendar to avoid confusion.

and took over the power in Ethiopia, leading to twenty years of communist regime. During the hard communist time, Ethiopia had a series of famines that resulted in the dead of one million people (Ibid). In 1994, Ethiopia had its first multiparty election. In August 2012, the Prime Minister Meles Zenawi died unexpectedly. His position is taken over by Hailemariam Desalegn without major problems (Ibid). Another major issue was the border dispute with Eritrea, resulting in a war between the two countries between 1998 and 2000. The war ended in the independence of Eritrea, although the borders are still secured by UN military forces. In 2011, Ethiopia experienced another major drought, after missing two rainy seasons. The government is now trying to make the country more resistant against droughts, by expanding irrigation practices and using the available water sources (FDRE, 2010).

After the communist period of rule, Ethiopia transformed to a Federal Republic with a constitution and democratic elections every five years. The federal government is responsible for national defence, foreign relations and general policy. The country has nine regional states based on ethnicity and two independent city states. The regional states are autonomous and are headed by a state president elected by the state council and have their own constitution, flag, anthem, language, capital city, executive administration and elected assembly (FDRE, 2013/Arsano, 2010, p.6). In the regional states, the woreda's (districts) are the lowest level of the government. In woreda's, there are also smaller entities called kebelles, but these do not have their own government offices (Edossa et al., 2007/Van Koppen et al., 2009). The Ethiopian government has a commitment to make the formal planning system decentralised, integrated and bottom-up. In practice, this transition appears to be difficult with shortages of finance and human capacity on government level and the practice shows still a sectorial planning (GW-MATE, 2011).

2.1 Water sector in Ethiopia

Ethiopia is endowed with an abundant water resource potential from rainfall and other sources (Awulachew, 2010). This potential is characterised by uneven distribution both in time and space. The great diversity of terrain has led to wide variations in climate, soils, natural vegetation and settlement patterns. Moreover, the water resources are severely underdeveloped and under-utilized (Edossa et al., 2007). Around 80% of the yearly rainfall is concentrated in the main rainy season from mid-June until mid-September. There is a short rainy season around March. The rest of the year is generally dry (Van Koppen et al., 2009/Awulachew, 2010). Depending on the altitude, the average precipitation per year is between less than 100 mm to more than 2.000 mm. There are areas with major lakes and rivers, and dry areas with deep groundwater tables (Van Koppen et al., 2009). Due to a lack in storage capacity, large spatial and temporal variations and low development of irrigation systems, most farmers are only able to produce one time a year. Moreover, frequent dry spells and droughts are putting more pressure on the agricultural production and are leading to food insecurity (Ibid). There is an average variation in rainfall of 30% year over year (Awulachew, 2010). There is however potential to irrigate more areas to secure the agricultural production.

Besides great diversity in climate, the increasing developments in irrigation and the growing awareness of the farmers of the importance of irrigation for their livelihoods are putting more pressure on the existing water resources, as well as developments in hydropower and industrialisation and the immense population growth (Awulachew, 2010). Water scarcity and degradation of its quality have therefore become the potential sources of conflicts in some parts of the country, but there are also conflicts arising without factors of degradation and scarcity. Moreover, water resources are managed in sectorial and fragmented approaches which led to increasing competition for water within sectors and within society (GWP, 2010). These developments make Ethiopia an interesting country to conduct a research on water-related conflicts on local scale.

2.2 Water Resource Management Policies

The government of Ethiopia, especially the Ministry of Water and Energy[2] developed different policies, strategies and regulations for the management of the country's water resources in order to eradicate poverty. Below, the important policies for irrigation development are discussed, starting by the national five-year plan, followed by the water resource management policy, strategy and water sector development programme. The descriptions are focused on issues relevant for irrigation developments, but the policies itself are directed to management of all water resources and purposes.

2.2.1 The Growth and Transformation Programme

The government of Ethiopia is trying to improve the current water supply and irrigation situation, by stimulating water-centred development; an approach where water resource development is being integrated with economic development and land-use planning (GW-MATE, 2011). Ambitious goals have been formulated for agriculture-based industrialisation and improving access to basic services. These goals are part of the Growth and Transformation Programme (GTP), a five-year development plan for eradication of poverty and broad-based accelerated and sustained economic growth (MoFED, 2010b). The GTP plan is developed for the period 2010/11-2014/15. The long term vision of Ethiopia is *"to become a country where democratic rule, good-governance and social justice reign, upon the involvement and free will of its people, and once extricating itself from poverty to reach the level of a middle-income economy as of 2020-2023"* (MoFED, 2010b, p.21). The plan describes goals for the different (economic) sectors in Ethiopia.

For agriculture, the goal is to develop a modern and productive agricultural sector using enhanced technology. There need to be a shift to production of high-value crops and best practices of model farmers will be scaled up for use by other farmers (MoFED, 2010b). Besides attention for the development of medium and large scale irrigation systems (from 2.6% in 2009/10 to 15.6% in 2014/15), small-scale irrigation will be the main priority (Ibid, p.36). The plan wants to integrate agricultural developments with rural development. The main goals for small-scale agricultural developments are: scaling up best practices from model farmers; expanding irrigation developments and improve natural resource conservation; and stimulate the production of high value crops (Ibid, p.45). The target for small scale irrigation is to develop 1850 million hectares irrigated land by 2014/15 using small scale systems (the baseline is 853 million hectares in 2009/10) (Ibid, p.48). The developments in small scale irrigation will generate more income for the small-scale farmers and make them more resistant against climate change.

2.2.2 Ethiopian Water Resource Management Policy

The water resources are governed by the 1999 Ethiopian Water Resource Management Policy. The overall goal of that policy is: *"to enhance and promote all national efforts towards the efficient, equitable and optimum utilization of the available Water Resources of Ethiopia for significant socioeconomic development on sustainable basis"* (MoWR, 1999, p.1). One of the main principles is that water is a natural endowment commonly owned by all the people of Ethiopia. And the government wants to develop the country's water resources for economic and social benefits of the

---

[2] The Ministry of Water and Energy exists since 2010, before it was the Ministry of Water Resources. The policy documents are from before 2010 and therefore refer to MoWR (Ministry of Water Resources).

people, on equitable and sustainable basis (Ibid). Water is seen as an economic and social good. The policy also considers water resource development, utilization, protection and conservation as connected which needs an integrated approach for the best result. The involvement of different stakeholders is needed for effective water resource management and the community is supported in management and initiatives related to water developments (Ibid). Furthermore, the policy is encouraging the development of a coherent and streamlined institutional framework for the management of water supply at Federal, Regional, Woreda and Kebelle levels and clearly defines the relationships and interactions among them (Ibid, p.29).

The policy is recognising the importance of water developments in agriculture for the rural population, since 80% of people's livelihoods is based on farming and livestock agriculture. Water developments in agriculture can increase the productivity of the rural population. The policy is also stressing the need to increase the productivity for meeting the needs for the growing population. The specific objective for irrigation is: *"to develop the huge irrigated agricultural potential for the production of food crops and raw materials needed for agro industries, on efficient and sustainable basis and without degrading the fertility of production fields and water resources base"* (MoWR, 1999, p.35). Irrigation is seen as integral part of the water sector. The main objectives of the irrigation policy part of the Ethiopian Water Resource Management Policy are (Ibid):

1. Development and enhancement of small scale irrigated agriculture and grazing lands for food sell-sufficiency at household level.
2. Development and enhancement of small-, medium- and large-scale irrigated agriculture for food security and food self-sufficiency at national level including export earnings and to satisfy local agro-industrial demands.
3. Promotion of irrigation study, planning and implementation on economically viable, socially equitable, technical efficient, environmental sound basis as well as development of sustainable, productive and affordable irrigation farms.
4. Promotion of water use efficiency, control of wastage, protection of irrigation structures and appropriate drainage systems
5. Ensuring that small-, medium-, and large-scale irrigation potential projects are studied and designed to a stage ready for immediate implementation by private and/or government at any time.

The involvement of different stakeholders is also encouraged by the policy. There is a special focus on the involvement of the local users, especially female users. When projects are designed, there should be attention for identification of all the relevant stakeholders and room for these stakeholders to consult with each other and discuss issues around the water systems. The government should pay attention for a legal basis to ensure an active and meaningful participation of all stakeholders (Ibid, p.35). In this way, the management of water systems will become more sustainable.

2.2.3 Ethiopian Water Sector Strategy

To make sure the Water Resource Management Policy will be implemented, the Ethiopian Water Sector Strategy was developed in 2001. The objectives of the strategy are similar to the objectives in the Water Resource Management Strategy. One of the main guiding principles is that water resource development will be rural-centred, decentralised managed and developed from a participatory approach (MoWR, 2001, p.2). An important action is the assessment and development of the country's surface water resources and groundwater resources, in order to increase the drinking water and irrigation coverage. The highest priority is the water allocation for drinking and sanitation purposes, followed by water requirements for livestock. In multi-purpose plans, these principles will be taken into account (Ibid). Part of the overall Water Sector Strategy is the Irrigation Development Strategy. When the strategy was written, the irrigation developments in the country were negligible (estimated at 197,225 hectares). It is recognised that irrigation development is the key to sustainable agricultural development (Ibid). The main objective of this sub-strategy is: *"to exploit the agricultural production potential of the country to achieve food self-sufficiency at the national level, including export earnings, and to satisfy the raw material demand of local industries, but without degrading the fertility and productivity of country's land and water resources base"* (MoWR, 2001, p.23). For the development of irrigation systems, it is important to take into account technical, environmental, financial, economic, social and institutional aspects.

2.2.4 Water Sector Development Programme
In 2002, the Ministry of Water Resources (MoWR) formulated the Water Sector Development Programme. This programme describes the development plans in the water sector for the period 2002-2016. The goals for drinking water and irrigation are (MoFED, 2010b, p.15):

1. Making clean drinking water available to the larger segments of the society.
2. Making water available for livestock in critical areas such as the nomadic areas.
3. Increase medium and large scale irrigation development so as to increase agricultural products and ensure food security.

In this plan, priority projects have been identified with a detailed implementation plan. The programme is developed with assistance of consultancy companies and the United Nations Department of Economic and Social Affairs (Ibid).

According to the Water Sector Development Program, Ethiopia has to following main problems in the water sector (MoWR, 2002, p.20-21): low institutional capacity and effectiveness; shortage of financial capacity; lack of coordination between the various implementation institutions (e.g. federal government, regions, districts and NGOs); lack of effective technologies and equipment; absence of involvement of all relevant actors; lack of data and information about the water sector; and low water efficiencies. All these problems are keeping the government from providing water to all its citizens. Although it is good the government had identified these problems, the question is whether there can change the situation. In a recent book by Kebede (2013), these problems are still described as leading to lack of sustainability and reliability of rural water systems in Ethiopia. Also constraints like corruption, low sense of ownership and inadequate communication between the different stakeholders and institutions is mentioned as source of problems (Kebede, 2013).

2.3 Irrigation systems and developments
This research is focused on small scale irrigation schemes, which is a policy priority in Ethiopia for rural poverty alleviation and growth (Tucker and Yirgu, 2010)[3]. Variations in rainfall between and within years make rain-fed agriculture a gamble with nature (Sreeramulu, 1998). In Ethiopia, the availability of water is crucial for the food security and income of local farmers (Van Koppen et al., 2009). Developments in irrigation make it possible for farmers to produce more times per year and supplement during unreliable rain seasons. In this way, small scale irrigation can promote rural food security, poverty alleviation and adaptation to climate change. When households are using irrigation, this means an average 20% increase in annual income, in some cases up to 300%, due to cultivation of higher value crops, intensified production and reduced losses. Nutrition was said to have improved as

---

[3] Small-scale irrigation < 200 ha, medium-scale between 200 and 3.000, and large-scale > 3.000 (interview 6).

various fruits and vegetables became locally available (Tucker and Yirgu, 2010). The irrigation systems can improve the agricultural sector in general, resulting in more job opportunities and possible more export of agricultural products (Awulachew, 2010). Irrigation can have different water resources, such as groundwater or surface water. Groundwater is a dependable source of irrigation for small areas (Sreeramulu, 1998).

Different kind of systems can be used for irrigation, depending on the location, the water source and other contextual factors. For surface irrigation, the water is distributed by gravity to the fields using furrows, basins, canals, etc. These systems are often not very efficient due to evaporation and soil infiltration (Sreeramulu, 1998). Also other climate factors, like wind and temperature have an influence on the water availability in these open-air systems. In river, systems of check dams, diversions and small dams can store water in reservoirs and lead it to the fields. These systems can irrigate small areas, but also cover larger areas. The water can be transported in modern canals (concrete or plastic pipes) or traditional hand dug canals (Ibid).

Drip and sprinkler system are considered as modern irrigation systems, which are more efficient than the traditional one, but also more expensive in construction and maintenance. In these systems, deep boreholes are used to pump groundwater to the fields. In drip systems, water is daily applied near the root zones of the crops. In this way, the water use is considerable reduced. The water system is treating the water before usage, to make sure the water is free of sand and other particles which can clog the drip system (Sreeramulu, 1998). Sprinkler systems consist of a network of pipes and sprinkler pipes which can simulate rain situations (Ibid). Most of the time, the drip and sprinkler systems are combined. Some deep boreholes are constructed without drip and sprinkler systems; in these cases water is distributed via canals (hand-dug or lined) or via a tank system in various part of the irrigated area. In areas of shallow groundwater tables, hand dug wells can also be used to access the groundwater for irrigation development. The pictures on the next page illustrate some of the different irrigation schemes visited in Ethiopia.

The potential of irrigation for rural development in Ethiopia is high, but the developments are going slow. Ethiopia had 4-5% irrigated of the land under cultivation, which equals around 640,000 hectares in 2010 (Awulachew, 2010, p.16). The potential land what can be irrigated with the current sources is over five million hectares, which could improve the food security for up to six million households (Ibid). There are however some major challenges delaying the development of irrigation in Ethiopia. According to a study of the International Water Management Institute (IWMI) the three major constraints causing a discrepancy between irrigation plans and delivery are: institutional capacity and capability, technical capacity and tools and inadequate policies and regulations (Ibid). Other weak points of irrigation developments in Ethiopia are: low levels of efficiency and performances of existing systems; lack of finance; inadequate monitoring; weak extension services; and low protection on the sustainability of irrigation development (Tafesse, 2003). Another weakness is the low involvement of women in irrigation management. Often, women work on the land, besides working in and around the house, but they are not part of water committees focused on irrigation (Ibid).

Figure 2.2: examples of irrigation systems.

Explanation pictures:
A: Deep borehole with hand dug canals to distribute the water to the land, in Alamata woreda.
B: Hand dug well with motor pump to transport the water from the well to the land, in Alamata woreda.
C: A river diversion with lined canals for water transportation, in Raya-Azebo woreda.
D: Deep borehole with tank and plastic pipe system, in Kobo woreda (the picture shows one of the tanks).
E: Example of a drip pipe with one of the holes, in Alamata woreda.
F: Three sprinkler in the fields, in Raya-Azebo woreda.

3. Theoretical framework

> *"Water-related conflicts can only be dealt with through effective water governance" (Ravnborg, 2004, p.5).*

The previous chapters presented the importance of irrigation developments for the rural livelihoods in Ethiopia. However, the development of irrigation system can also have negative consequences and possibly lead to situations of conflicts between farmers. Before the empirical cases are presented, first the theoretical side will be discussed, looking at the work of other scientist on this subject and with attention for defining of some key concepts used in this research. In line with the research questions, attention will be paid to water conflicts and cooperation, institutions and conflict management. When discussing water conflicts and cooperation, different factors contributing to the management situations will be discussed. The section about institutions elaborates on some definitions of institutions, resulting in a definition used in this research. The section about conflict management discusses approaches which can be used to solve conflicts. This chapter will end with a framework bringing the different concepts together.

3.1 Water conflicts and cooperation

There are different theories that describe pathways of conflict and cooperation in natural resource management. Access to natural resources is important for the livelihoods of people, especially in rural parts of Africa where people depend on agriculture for their income (Ravnborg et al., 2012). Resource conflicts can jeopardise the access of the rural people to the natural resources and on the functionality of irrigation systems, which can have a negative influence on the lives of the rural poor.

Resource scarcity used to be emphasized as a cause for natural resource conflicts. Resource scarcity is caused by insufficient supply, increasing demand, unequal distribution or resource depletion and degradation (Homer-Dixon, 1999). The Development Assistance Committee of the OECD distinguishes two types of resource scarcity; physical and economic scarcity. Countries experience physical water scarcity when there is not enough primarily available water to produce enough food for themselves (e.g. in North Africa and the Middle East). Countries with economic water scarcity have enough primarily, naturally available water, but they lack the infrastructure and institutions to make use of it (OECD, 2005, p.5). According to a study of Homer-Dixon (1999), scarcity of renewable resources can contribute to civil violence, including insurgencies and ethnic clashes. However, scarcity alone if often not the cause for conflict; scarcity is always interacting with other economic, political and social factors to produce effects on the society (Ibid, p.177). Boserup suggests that scarcity can be overcome by technological innovation, efficiency, conservation or other forms of human ingenuity and cooperation (Boserup, 1965). In that sense, environmental scarcity is not always a bad thing; it can stimulate technological entrepreneurship and institutional change (Homer-Dixon, 1999).

Besides the focus on scarcity, other explanations are given in scientific studies on natural resource conflicts. One school of thought is explaining resource conflicts with management of the natural resources; rather than water scarcity in itself, water-related conflicts are caused by the way in which the water systems are managed. Governing water inevitably means governing conflicting interests, because all facets of society are depending on water sources (Ravnborg, 2004, p.8). The way water institutions are dealing with the conflicting interests and are functioning can influence the local situation, whether a conflict will arise or not. Ostrom (1990) suggests that resource related conflicts are caused by poorly defined and/or poorly governed property, in other words, by institutional failure. Institutional and social arrangements within society are the key determinants of prosperity and not the availability of natural resources (Homer-Dixon, 1999). Ostrom is also describing that cooperation will most likely happen when the resource is neither abundant nor very scarce (1990).

The way irrigation systems are managed (functioning of institutions) is also influencing other factors mentioned in literature about water-related conflicts. Recent studies stress the role of distribution and access issues. Resource conflicts are often related to inequality and distribution issues within rural societies and local management institutions have an important role in managing this distribution between the different stakeholders (Cold-Ravnkilde, 2012). However, the allocation of water can be a source of tensions (OECD, 2005). Water is not evenly distributed in time and space and does not recognise administrative boundaries (OECD, 2005). As a result, management of water resources equals conflicting interests and distribution/ access issue can result in tensions.

Due to an increase in competition for water resources, there is a fear of an increased occurrence of local water conflicts (Ohlson, 2000; Ravnborg et al., 2012). However, for many rural communities in the South, conflict and cooperation over water has always been a fact of life, due to prevailing natural conditions and unequal patterns of distribution (Funder et al., 2012). These local conflicts often remain outside the attention of the authorities and other social actors outside the events and the involved parties (Ibid; Funder et al., 2010). Water related conflicts are issue based and diverse and they are changing over time as function of changing demands and options for water use (Ravnborg, 2004, p.13). Also, changes in water usage and new users coming in (e.g. large-scale farming and hydropower) have an influence on the local competition for water resources (Ibid). This competition between different types of uses can be called multiple-use conflicts, which can be defined as *"systems that allow efficient and effective supply of water from different sources to communities for their domestic and for their productive purposes and that allow interaction with providers of water related services"* (Penning de Vries, 2007, p.79). Local institutions are again important here to guide the usage of a water resource.

Social conflict is not always a bad thing. Conflict situations can be considered as normal elements in social interactions and can be a transition to a situation of better balanced resource management and usage (Van Steenbergen, 2011/Constantinos, 1998). Moreover, useful change in the institutions and processes of governance can be achieved (Homer-Dixon, 1999). There is however a difference between small conflicts (more disputes and disagreements) and escalating conflicts with negative influence on the community. Not all conflicts are normal or good. On the other hand, cooperation is not always the ideal situation. Some situations of cooperation are based on inequality or cause resource degradation (Van Steenbergen, 2011). Besides conflict and cooperation, there is a third situation which is probably more common than conflict and cooperation, a situation of increased resource use, leading to scarcity and the undermining of the resource base without anything happening. These situations may be called 'void' and characterises a situation of increased pressure on the natural resources without conflict or cooperation being triggered (Ibid). Conflict in fact may be better because it would at least create the basis to set in place a system of active and fair resource management.

One may assume that groundwater is more amenable to protracted void situations. Because groundwater is invisible and because aquifer boundaries are not well defined, overuse and resource degradation can take place without anyone taking notice or understanding the dynamics. Also defining access or user rights is more difficult in groundwater management than in divisible resources such as land or surface water. Conflicts are less likely to happen, but at the same time access to groundwater can be influenced by local power play and political issues (Van Steenbergen, 2011). This local power play and political issues are however also present in surface water situations. Water management can be considered as highly complex and political (Carius et al., 2004). Water conflicts can be seen as social processes that reflect social, political and historical struggles in rural societies (Cold-Ravnkilde, 2012). At the same time, political, socio-economic and cultural factors on different levels determine whether tensions around water systems lead to conflict (OECD, 2005).

In this discussion about water conflicts and cooperation, different explanations have been given for resource conflicts. These explanations are translated to factors to serve as tool for analysing management situations in Ethiopia. The first factor is scarcity, which is traditionally seen as explanation for resource conflicts. This factor has been divided into economic and physical scarcity. A second important factor the functioning of institutions, related to water governance. The way in which local institutions related to water are functioning determines the nature of the local management situation, whether conflict or cooperation will arise. The functioning of institutions is also influencing the other factors, except physical scarcity and therefore is an important factor in water management. A third factor is the distribution and access to water. Since water is unevenly distributed and represents conflicting interests, the way water is distributed and who has access to it, is important for the occurrence of disputes and conflicts. A good distribution schedule made by e.g. a local water committee can deal with economic scarcity and in that way prevent local conflicts. This factor is related to the next factor; competition. With this factor, competition between different types of uses is meant, with other words multiple-use issues. A last factor is the social and political relations within the community. Africa is a continent where social relations and family are very important and local politics can influence the live in communities. The ways in which local water sources are managed are related to the social relations and norms of behaviour and the local politics. Most of the time not one factor responsible, but a combination of factors, interrelated in a big and complex interaction and dynamics (Carius et al., 2004). These factors are represented in the theoretical framework at the end of the chapter.

Besides factors leading to conflicts or cooperation situations, water related conflicts can also be classified in different types of situations. The main categories mentioned in the literature are intra- or inter-community and upstream-downstream conflicts (Ravnborg, 2004). Most conflicts are happening on local scale and are affecting people locally, within one village or between two villages (for example when they share a water system). Upstream-downstream conflicts are common in rivers where different communities are using the river for various purposes. Most of the conflicts are related to communal/publically owned water sources (Funder et al., 2010). In cases of rivers, there is also the possibility for multiple-use conflicts, where there is for example not enough water for domestic and productive use of the water (Ibid).

3.2 Institutions

The discussion about factors causing situations of conflict and cooperation revealed the importance of good working institutions, but before we can say something about institutions, we first need a clear definition. The term "institution" is a widely used and can describe rules that structure social interactions, or organisations who implement these rules. Ostrom defines institutions as *"the set of working rules that are used to determine who is eligible to make decisions in some arena, what actions are allowed or constrained, what aggregation rules will be used, what procedures must be followed, what information must or must not be provided, and what payoffs will be assigned to individuals dependent on their actions"* (1990, p.51). This definition is focused on the rules institutions represent to order social life. Cold-Ravnkilde uses a narrow definition of institutions in her PhD research about water conflicts: *"institutions are both state and non-state forms of (social) organisations that are held together by and produce rules, norms and practices"* (2012, p.18). This focus on organisations is useful for this thesis, but the rules that structure social interaction also deserve a place in the definitions. Hodgson defines institutions as *"systems of established and embedded social rules that structure social interactions"* (2006, p.8). Organisations are defined as *"special institutions that involve (a) criteria to establish their boundaries and to distinguish their members from non-members, (b) principles of sovereignty concerning who is in charge, and (c) chains of command delineating responsibilities within the organisation"* (Hodgson, 2006, p.8). Hodgson's definition of institutions and organisations is closely related with the purpose of the thesis and will be used as basis definition.

One common way of approaching institutions in natural resource management is via the concept of common pool resources (CPR). This concept is underlying decentralised natural resource management initiatives and the idea that local institutions can be crafted to manage natural resource in a sustainable way (Cold-Ravnkilde, 2012). Elinor Ostrom defines common pool resources as: *"a natural or man-made resource system that is sufficiently large as to make it costly (but not impossible) to exclude potential beneficiaries from obtaining benefits from its use"* (1990, p.30). In her analysis, she describes that the management of common pool resources often fails; models with the market and the state do not provide effective management solutions. Ostrom suggests that common pool resources should be managed by long-enduring, self-organised and self-governed institutions. Although the CPR concept is well known and widely cited, there are also some critiques. The theory sees people as self-interested rational agents, but fails to recognize that people's concerns, interests and motives can overlap and change over time (Cold-Ravnkilde, 2012).

Defining institutions is a first step in discussing institutions. In the definition of institutions and organisations, the focus lays often on formal/government institutions. There are different government organisations dealing with water, from the ministry of water on national level, to regional water board, or local water departments. The main tasks of these governmental organisations are dealing with data, information and knowledge about water systems and support local management entities (Kebede, 2013). In many countries, the government is the owner of land and can decide to lease land to investors. Various studies proved the fact that water management is failing for many reasons. The most important reasons for Africa are lack of adequate water institutions, inadequate administrative capacity, lack of transparency, ambiguous jurisdiction, overlapping functions, fragmented institutional structures and lack of necessary infrastructures (Carius et al., 2004, p.61). To be successful, water management need to consider cultural norms, societal rules, values of people, behaviours of different stakeholders (Wengert, 1983).

Water management is equal to managing conflicting interest and is therefore highly complex. Besides formal institutions, informal institutions and customary laws also play an important role and this is less prominent present in the various definitions. In Africa, formal institutions tend to overshadow the local informal ones although the latter guide day-to-day interactions on water use (Sokile and Van Koppen, 2004). Also other aspects of rural life in Africa are more guided by informal traditional organisations and rules than the formal governmental institutions. Several studies have acknowledged the fact that informal local level institutions can make a difference in water management (Ibid). In sub-Saharan Africa, water is a basis for life for agro- and for pastoral societies and is anchored in a deep socio-cultural and economic context. Research showed that local water arrangements are more efficient, more cost-effective, longer-lasting and more widely accepted among local users than most top-down state-driven institutions (Ibid). Greater attention need to be paid to the existing and potential role of local political systems and informal networks in sustainable form of water management (Turner, 1999). However, understanding these informal organisations and networks will be more difficult than their formal counterparts.

This thesis will focus on the role of different types of organisations, which are part of the larger concept of institutions. Although the role of (government) organisations will be the main focus in the analysis on institutions, policies and rules that structure behaviour around water systems will also be discussed, as the role of informal organisations and rules. To avoid confusion between the terms institutions and organisations, the term institutions is been divided in different categories:

- *Government organisation*: on different levels, Ministry of Water and Energy, regional water bureaus, woreda water bureaus etc.
- *Government policies*; regulations, procedures on different levels.
- *Local organisations*: water committees, informal organisations involved in water management
- *Local rules and regulations*: traditional, informal rules on local level.

3.3 Conflict management

Water-related conflicts are often caused by the way in which water and its use are managed (Ravnborg, 2004). Conflict and tensions, but also situations of void can hamper the development of water systems and result in degradation or over-use of the water source and jeopardise the sustainability of the systems. Functionality and sustainability are closely related: good sustainability keeps functionality rates up and vice versa. WaterAid considers a water point sustainable when all the necessary components that keep a water point functional are in place – i.e. if the technology, management, finances, technical expertise, availability of spare parts, dependable water source, etc. (WaterAid, 2009, p.1). Effective water governance is needed to ensure sustainable water management. Water governance can be defined as *"the range of political, social, economic and administrative systems that are in place to develop and manage water resources, and the delivery of water services, at different levels of society"* (Ibid, p.8). Conflict prevention and resolution is water-resource management is a matter of recognising and understanding conflicting interests relating to water governance at different levels (Ibid, p.9). However, formal mechanisms for conflict resolution are still rare in local management practices and local rules and regulations.

Conflicts can be solved by using different conflict management tools to anticipate, prevent and react to conflicts (GWP, 2010, p.2). A first step in resolving conflicts is a proper identification and analysis of the conflicting issues relating to water causing the conflict (Ravnborg, 2004). The most important conflict management tools are consensus building, decision support and modelling tools, decentralised participatory multi-stakeholder platforms, and interventions by organisations (GWP, 2010). The involvement of all stakeholders is important for a successful approach. A solid conflict management strategy will involve a combination of tools used to induce the parties to open up, identify the real issues behind the publicly pronounced positions and find out "win-win" solutions that leave both the parties better off with the outcome (Ibid). Also indigenous knowledge and processes should be used, when they are useful for the situation (Ravnborg, 2004). It is important that there is attention for conflict prevention and resolution during formulation of policies and regulatory frameworks, so not only in case of actual conflict or disputes. It is however difficult to make a blueprint for conflict resolution and prevention, since conflicts are issue-bases and diverse, and they can change over time related to changes in water use (Ibid).

Most water-related conflicts are issue based and change over time related to changing demands and user-intensity. Therefore, a universal way of solving conflict will not work (Ravnborg,

2004). Although stakeholder participation sounds like a good option for conflict resolution, this should be done carefully. There should be an environment of trust where different stakeholders can speak freely without escalating the conflict or reinforcing existing power balances. Who is mediating in a conflict needs to be careful considerate. The water committees should also be trained in conflict resolution and local stakeholder participation (Ibid). Moreover, local management of water resources needs to be combined with an analysis of livelihood strategies, land-use pattern and access to markets, because irrigation developments are closely related to these developments as well.

In Africa, water committees are often established for the management of local water systems (Harvey and Reed, 2006). After construction, the government is transferring the management of the systems to local water committees who are responsible for the daily management and operation. Although the establishment of local water committees is seen as the way to manage local water systems sustainably, they are not always successful (Kebede, 2013). The actual stakeholder participation in the committees as a whole (including the formulation and renegotiation of the policy, legal and regulatory frameworks) has been limited. And water committees have often limited and/or unclear mandates and tend to reproduce existing power balances (Ravnborg, 2004). Therefore, more attention is needed to the community management of water systems to make them more sustainable and resistant to conflict situations.

On international level, there have been discussions over water management for a long period of time. One of the results of the discussions is the concept of Integrated Water Resource Management (IWRM), an integrated approach to manage water resource on national scale. IWRM is recognising the multiple interest, -uses and -stakeholders involved in water management (Funder et al., 2010). To ensure water security, different challenges have to be dealt with: meeting basic needs, food security, protecting ecosystems, sharing water resources, managing risks, valuing water and governing water wisely (Savenije and Van der Zaag, 2008). IWRM acknowledges the entire water cycle with all its natural aspects, as well as the interests of the water users in the different sectors of society (Savenije and Van der Zaag, 2008, p.292). Different countries in Africa are trying to implement principles of IWRM in their water management, to improve water management and be prepared to deal with problems (Ibid). Ethiopia also has done attempts to implement the principles of IWRM in water policies. Most clearly in a pilot project of the Ethiopia Country Water Partnership, part of the Global Water Partnership. This project wants to promote IWRM in Ethiopia and more general, promote sustainable water management (ECWP, 2007).

3.4 Research framework

This chapter discussed the main theories related to the research. The main concepts are combined in a framework. For the research, the different management situations are defined as follows:

- Conflict: open competition for the water resource, where conflicts and tensions are arising between users, managing institutions or the (local) government. The competition is leading to a negative influence on the water system, but does not necessarily have to be violent.
- Void: situations where there is no conflict or cooperation, but the water resource can be deteriorated or developed blindly, which can jeopardise the sustainability of the system.
- Cooperation: development of access systems, laws, institutions and monitoring systems. In this way, different stakeholders share the water access.

These situations are central in the research framework and are seen as dynamic. From the literature, different factors have been identified that influence the management situations; economic- and physical scarcity, distribution/access, functioning of institutions, competition and social relations within the community. With functioning of institutions is meant the broad definition of institutions, with other words, systems of established and embedded social rules that structure social interactions. This can be policies of the government, government organisations, local management organisations and local rules and regulations. The functioning of institutions also has influence on some of the other factors. This thesis will look whether these factors are indeed influencing the different situations. The thesis also analyses the cases where conflict resolution and conflict prevention have been used, these elements should have a positive influence on the situations and guide them towards cooperation. Institutions are involved in conflict prevention and resolution, therefore there is a connection between these elements.

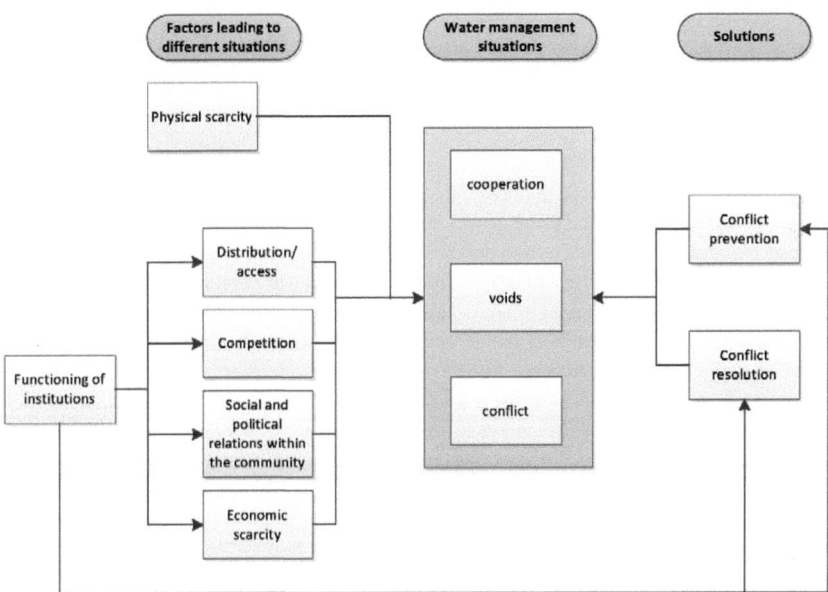

Figure 3.1: conceptual framework

4. Methodology

The strength of a geographer is the preoccupation with places, which allows him to relate to different settings, and compare and understand the factors and driving forces behind the differences (Van Steenbergen and Verheijen, 2006, p.283).

Conflicts can be considered as a normal element in social interactions (Costantinos, 1998). Despite this fact, conflicts are often complex processes, which can be sensitive and difficult to understand for an outsider (Goodhand et al., 2002). This makes researching conflicts a challenging process. This chapter will describe the approach chosen for this research, the data collection method and the limitations of the research. The chapter ends with a table indicating the visited places and organisations.

4.1 Approach

Goodhand et al. developed a manual for conducting conflict assessment, further called the DFID manual (UK Department for International Development). The authors make the note that the method they describe needs to be adapted to the specific situation, since every situation is different and may need an altered approach (Goodhand et al., 2002, p.7). Before the conflicts can be solved, a systematically analysis of the conflict and the dynamics around it needs to be conducted. To be able to compare the different cases in Ethiopia, a research approach was developed, based on the DFID manual and on methodologies used in articles about local water-related conflicts (Ravnborg et al., 2012/GWP, 2010). Although the DFID manual is developed for violent conflicts on larger scale, the basis of the conflict analysis could be used for this research.

According to the DFID manual there are three key stages in conflict assessment; analysis of conflict, analysis of international responses to conflict and development of strategies and options (Goodhand et al., 2002, p.10-16). The first stage (analysis of conflict) is most relevant for this research. The second and third stages are more directed towards actions of international actors and are therefore not used in the research as such. However, there is attention for responses from the different actors on the conflicts and how these local conflicts can be prevented and resolved. The first stage of conflict assessment is divided in three elements, namely structures, actors and dynamics, which should be viewed holistically (Ibid). When assessing the structural factors behind the conflict, it is important to perform a contextual analysis (based on historical, physical and demographic features), identify the sources of tensions and make a combination of the two analyses (Ibid). The second important element is the assessments of important actors who influence or who are affected by the conflict (Ibid). An assessment of actors should be based on identifying the relevant actors and determine per actor the interests, relations, capacities and incentives. The third element of dynamics looks at the likelihood for conflict situation to increase, decrease or remain stable, especially the long term trends and short term triggers that influence the conflict. Institutions and processes that can mitigate or manage tensions and conflicts are important to assess in this part (Ibid). These different elements are visualised in figure 4.1.

TABLE 1: CONFLICT ANALYSIS		
(i) Structures	(ii) Actors	(iii) Dynamics
Analysis of long term factors underlying conflict:	Analysis of conflict actors:	Analysis of:
• Security • Political • Economic • Social	• Interests • Relations • Capacities • Peace agendas • Incentives	• Long term trends of conflict • Triggers for increased violence • Capacities for managing conflict • Likely future conflict scenarios

Figure 4.1: Three elements of conflict analysis in the DFID manual (Goodhand et al., 2002, p.10).

For the research, the three elements distinguished in conflict analysis (structures, actors and dynamics) are seen as connected but not as strict categories. As appeared from the theoretical framework, the role of institutions is identified as important in this research. Within the concept of institutions, organisations on government and community level have a prominent place. Relating this concept to the DFID framework, these organisations are part of the actor element, but are also determining the structures locally and influencing the dynamics of the situation. As described in the theory chapter, not only conflict situations were focus of the research, also situations of void and cooperation have been investigated and will be described in the next chapters. The focus of the research is therefore on water management situation, but also to understand these situations, structures, actors and dynamics are important elements to understand.

Water management situations are seen as dynamic situations that can change from nature. Conflict and cooperation are seen as two extreme situations, where a lot of situations can fall in between. Situations where people have a competing interest in water resources can lead to a conflict situation, but also to cooperation or void situations in which elements of both conflict and cooperation can return. From the fieldwork, a wide variety of cases has been chosen to illustrate this difference in the nature of situations. To describe water management situations, detailed information of the cases is needed. Therefore, claims and narratives of different stakeholders involved are important to capture and understand (Cold-Ravnkilde, 2012, p.38). The interviews are used anonymously, only location and date are mentioned. Also government officials are named by their function instead of their name (also in the interview list in table 4.2 and appendix II & III).

4.2 Data collection

The goal of this research is to reach a holistic understanding of complex processes of water management in Ethiopia, by describing and analysing empirical cases of water management situations in Ethiopia. In the preparation phase of the research, a literature review has been conducted to become familiar with theories about water management and local conflicts. A general research design was made, which was further developed in Ethiopia itself, after meetings with local professionals and supervisors. In Ethiopia, I identified the visited cases via informal networks of professionals working on the topic and based on the information at woreda level and with help of my local supervisors. This was needed, since most water-related events do not come to the attention of outside institutions (Ravnborg et al., 2012). The fieldwork has been conducted in February-May 2013.

To describe and analyse the different management situations, in-depth knowledge about the local cases is needed. In the interviews and field visits, I paid attention to reactions on conflict situations, the situation in the past and the way the various actors want to move forward to acquire better knowledge about the dynamics of the management situations. For the fieldwork, I needed a qualitative research design, because the research is dealing with sensitive and complex situation (Goodhand et al., 2002). I chose to use semi-structured interviews to be able to collect detailed information about the different cases. The semi-structured interviews were held with different types of actors; farmers using the irrigation systems, members of the water committee responsible for the management of the systems, woreda officials from the agricultural and water bureaus, regional officials from the agricultural and water offices, the Ministry of Water and Energy and NGOs working in the research field. I also tried to assess the role of traditional organisations and rules, since these are important in African cultures. The semi-structured question lists were developed with consultation of the article of Ravnborg et al., 2012, the DFID manual and the question lists in the appendix of the report of Arsano et al., 2010 (see appendix I). The question lists were used to guide the interviews and make sure the different cases were comparable, but also to create room to adapt the interviews to the specific situations. Ethiopian professors and my supervisors explained that it is not common for Ethiopians to talk about conflicts. When you would directly ask for conflicts, people would likely deny that there were problems. Therefore, the semi-structured questions were adapted and the word "conflict" was avoided in the interviews, especially in the beginning of the interviews.

Data has been collected in three regional states in Ethiopia: Tigray, Amhara and Oromia. In total, seven woreda's (districts) have been visited: Kilite-Awlalo, Atsbi-Womberta, Raya-Azebo and Alamata in Tigray regional state; Kombolcha and Kobo in Amhara regional state; and Degem in Oromia regional state (see map 4.2 for the locations). Moreover, I conducted interviews with regional and national government organisations in Mekelle, Bahir Dar and Addis Ababa. In total, around 60 interviews have been conducted next to various discussions with my local supervisors (see table 4.3 and appendix II & III). From all the data and visits seven cases have been selected. Especially in Raya-Azebo, Alamata and Kobo woreda, more water schemes have been visited, but the most interesting and outstanding cases have been selected, based on the aim to present a various group of cases and the available data for each of the cases. The experiences from the other cases have been used for the more general parts in the empirical chapters. The number of interviews per case differs from three to sixteen (see appendix II). In every case it was tried to talk to officials from the woreda water and agricultural bureau before visiting the schemes. In this way, a general impression of the dynamics in the woreda could be obtained. Only in Amhara region, the water bureaus were not as directly involved in irrigation schemes, therefore in these cases, only the agricultural offices were visited. Furthermore, in cases of systems with water committees responsible for the management, an interview was held with at least one member of this committee. In two cases, it was not possible to talk to other farmers, than members of the water committees.

The cases are not only interesting by themselves, but also in comparison with other cases. The intriguing issue in comparing Raya-Azebo and Alamata with Kobo is that they are located in connected valleys, but the level of development in the woreda's differs as well as the approach of development. The woreda's Kilite-Awlalo (sometimes called Wukro) and Atsbi-Womberta are interesting to visit, because they share a river and in the past there have been some tensions about the water availability and distribution. The woreda Degem is known for its competition over water between different kebelles. Kombolcha was visited to look at the involvement of a local NGO and their approach on promoting conflict management.

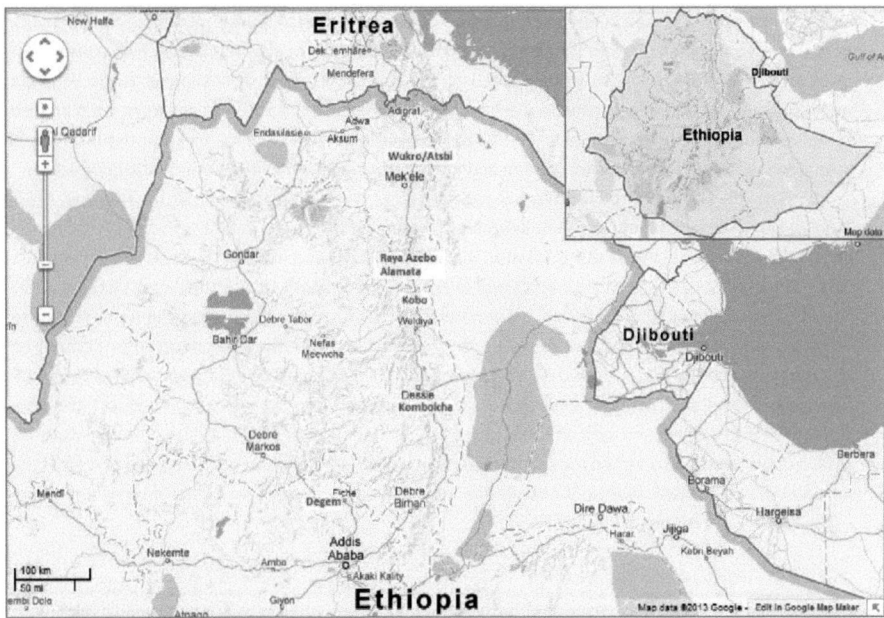

Map 4.2: Map of Ethiopia, with areas of visit in red (GoogleMaps, 2013)

As far as it was possible, background information about the visited areas has been collected before going into the field to get an impression of the background and structural factors present in management situations (e.g. local theses, woreda profiles, and feasibility reports of projects). The structural factors in each case were further gathered through interviews with local government officials and observations in the field. In the different woreda's, we first conducted interviews with woreda government officials from the water bureaus and the agricultural bureaus to get information about the situation in the woreda and the interesting locations for interviews with farmers and water committees within the woreda. For creating a complete picture of the situation, regional water bureaus and agricultural bureaus were visited in the regional capitals cities. Some regional/local NGO offices were visited, when they were involved in one of the cases or irrigation development in the area. In Mekelle and Addis Ababa, the universities have been visited and to discuss my research with some professors. The Ministry of Water and Energy was visited for access to relevant policy documents and more background information about the irrigation developments in Ethiopia. Also in the woreda and regional offices, access to data on irrigation developments was requested and most of the time given, but not all the data was relevant or comparable over different woreda's/regions. Data about cases of conflict were also requested, but this kind of information is not available in local offices at all. Most of the information about the water management situations was collected via semi-structured interviews with the relevant actors on different levels (e.g. farmers, water committees, woreda officials, regional officials and NGO's).

The field visits were conducted in three fieldwork parts. The five northern districts were part of the first part. This fieldwork was prepared by a few days in Mekelle (capital city of Tigray regional state), were arrangements were made with Mekelle University for transport and translation. A car with driver was available for ten days and Dr. Dessie Nedaw Habtemariam joined me during the fieldwork period to organise the visits and translate during the interviews. The plan was to visit five woreda's for interviews, therefore, the fieldwork needed to be concise. The car also allowed to do multiple interviews and visits in one day and to use the limited time efficiently. After the ten days, we managed to do field visits in all five woreda's and five days in Mekelle were used to work-out the interviews. Interviews with the regional water bureau and Relief Society of Tigray (REST)[4] were also conducted.

The second part of the fieldwork consisted of four days in Degem woreda. The first day, Asefa Kumsa, Taye Alemayehu and Dr. Seifu Kebede accompanied me to the field to show the woreda and arrange some practical things as a translator and a hotel for the coming days. The rest of the fieldwork was conducted by using public transport and walking to the villages to visit the water systems. After the first day alone with my translator, I got help with translation and visiting of the areas from two woreda water bureau officials as well. The third part of the fieldwork I conducted in Kombolcha woreda. There a local NGO assisted with transport and translation. As a result, I was able to finish the interviews soon. Bahir Dar, the capital city of Amhara regional state, was visited to conduct interviews with the regional water and agricultural bureau. The Amhara Design and Supervision Work Enterprise (ADSWE) office was also visited for an interview and background documents[5]. Between and after the different parts, I spent time in Addis Ababa to analyse my interviews. I also interviewed some organisations and experts in Addis Ababa.

4.3 Limitations of the research

Conflicts are sensitive issues, especially in the Ethiopia culture where it is not common to talk about conflicts. As a result, it was sometimes difficult to talk about the conflicts or tensions directly with the respondents. In the beginning of the interviews, the farmers or members from the water committee would mostly say there were no problems or conflicts. In some cases, after we won their trust, they would admit there were problems and explain them. Due to this fact, I expect that not all issues were discovered during the interviews. In the interviews with the woreda officials, trust was also an issue. In the beginning of the interviews, most officials were denying problems. When they went to the field with us, they became more open and explained more issues and became more open. The local knowledge from the woreda officials was very helpful during the fieldtrips to the irrigation schemes. However, bringing government officials to the interviews with water committees and farmers might have caused a situation where the real issues with the government might not have been discovered. The farmers might have thought we were doing the research for the government, since they accompanied us. Most of the time, the farmers were still telling complains about the government and it was also good to be able to ask the position of the government officials after the interviews.

Transport and money were issue that limited the research to some extent. Because of limited public transport and expensive private transport, the fieldwork had to be concentrated in a restricted number of days and observations of cases were limited. In this research is chosen for short and dense

---

[4] The REST is a local NGO who helps people in Tigray with relief programmes and food security.
[5] The ADSWE is a public organisation hired by the government to conduct study and design for water resource projects and oversee the construction of the projects (ADSWE, 2011).

visits, to be able to see more diversity in different places, in line with the explorative character of the research. A consequence is that not all cases had sufficient time to talk to all different stakeholders and observe the management situation. Ethiopia is a large country and most of the visited irrigation sites were difficult to use by public transport.

The research was conducted in three regional states were they speak different languages. Therefore, I was dependent upon different translators for most of my interviews. The interviews with the regional and federal government I was able to do in English, as well as interviews with NGO and research institutes. Most woreda officials were not confident enough to do the interview in English. During the first fieldwork period in Northern Ethiopia, we did most of the fieldwork in Tigray regional state where people speak Tigrinya and not always Amharic (the national language), especially the local farmers do not always speak Amharic in Tigray. Dr. Dessie is originally not from Tigray and does not speak Tigrinya. Therefore, for some interviews with farmers, we needed translation from the government officials who joined us to the field. In two interviews, part of the interview had to be translated from Tigrinya to Amharic to English, which likely caused information to be lost in translation. The interviews in Degem were held in Oromia (the local language) and translated to English.

One time (Degem woreda), I had a local translator from the woreda of the fieldwork. He had no experience in translating or conducting interviews and therefore it was sometimes difficult to explain to him what I needed to know. This might have impacted the quality of the data I collected. After one day, I got assistance of two woreda officials and this made the fieldwork a bit easier. This fact can also be an influence on the quality of data of the Degem case. But because we went two days without the woreda officials to the field, I still was able to see the dynamics between the community and the woreda government without presence of the woreda officials themselves, but it is a factor I had to take into account during the analysis of the data from Degem. Compared to the other cases, there was more time for observation in the Degem case, since I spent four days there conducting interviews for one case. Therefore I have enough data to present this case in this thesis. The other fieldwork periods, the translators were better skilled in speaking English. However, using a translator in semi structured interviews is still a limiting factor. Not all information is translated and the translator will translate the information on the way he is interpreting it.

In general, the research process went well. The respondents in Ethiopia were willing to take time for the interviews and provide most of the information needed for the research. Table 4.3 presents the different locations of the field visits and the organisation and actors consulted. I was also interest in documents about the woreda's, but the archives of especially the local government organisations were not always reliable or up to date. Therefore it was sometimes difficult to get comparable data about the different woreda's. Even data from the agricultural bureau and water bureau in the same woreda could differ (e.g. the number of hectares irrigated in the woreda). The collected data from local government officials was useful to get more background information from the woreda's, but the information from these documents was used carefully. And by visiting various organisations, it was tried to get the same information from different sources.

Table 4.3: Locations of data collection		
Woreda/city	**Regional state**	**Offices/actors interviewed**
Alamata woreda: *Kugitelemlem, Selamgelasi, Tumuga kebelles*	Tigray	- Water bureau - Agricultural bureau - Local REST office - World Vision office - Farmers/ water committees (six interviews)
Raya-Azebo woreda: *Tsigea, Genete, Worgeba kebelles*	Tigray	- Water bureau - Agricultural bureau - Local REST office - Farmers/ water committees (five interviews) - Minora business group
Kobo woreda: *Anoda, Jarota, Golina, Aybe kebelles*	Amhara	- KGVDP office - Agricultural bureau - Farmers/ water committees (six interviews)
Kilite-Awlalo woreda: *Haijalo, Agula, Abreha wa Atsbha*	Tigray	- Water bureau - Agricultural bureau - Kebelle leader Abreha wa Atsbha - Farmers (two interviews)
Atsbi-Womberta woreda: *Rufa falet*	Tigray	- Water bureau - Agricultural bureau - Farmers (three interviews)
Mekelle	Capital Tigray	- Regional Water Bureau - REST office - Mekelle university - Institute for Sustainable Development
Degem woreda: *Tumano, Ano Kare, Alidena kebelles*	Oromia	- Water bureau - Agricultural bureau - Farmers/ water committees (fifteen short interviews)
Kombolcha woreda: *08 and 09 kebelles*	Amhara	- Local Water Action Office - Agricultural bureau - Farmers/ water committee - Drinking water committee
Bahir Dar	Capital Amhara	- Regional Water Bureau - Regional Agricultural Bureau - Amhara Design and Supervision Work Enterprise (ADSWE)
Addis Ababa	Capital city, independent city state	- Addis Ababa university - Ministry of Water and Energy (three interviews) - Oromia Water Bureau - Water Action - Institute for Sustainable Development

5. Conflict, cooperation and void situations in Ethiopia

During the fieldwork period, different irrigation systems have been visited; some without having many problems, some experiencing tensions and conflicts. From the field visits and interviews, seven cases have been selected, which represent different dimensions of irrigation management situations and will be central in the next three chapters. This chapter will first introduce the seven cases from Ethiopia by means of a short description of the context and the problems. There is attention for the main characteristics, background and factors involved in the cases. In this way, the first sub-question, *what are the characteristics of local management situations and which factors contributed to the local management situations*, will be answered in this chapter. The next two chapters will discuss the management aspect and how conflicts are solved.

5.1 Description of the cases
The cases are presented in order from most violent situation to most cooperative situation, some of the cases are clear examples of conflict, some of cooperation, some have elements of both and fall in the category of void situations. In chapter 3, different factors have been discussed which, according to the literature, can cause situations of local water-related conflict: scarcity of water resources (economic and physical), distribution/access issues, functionality of institutions, competition for water (multiple use) and social/political relations within the community. In the discussion of each case, the factors involved in the case will be discussed.

      The first two cases are describing situations of conflict, where the first case the conflict has influence on the functionality of the system and in the second case, the conflict is more expressed in verbal disputes over the current situation, but still openly present in the daily management.

    a. Drip/sprinkler scheme Aybe, Kobo; conflict between farmers and the water committee

The Kobo valley is located in the North-Eastern part of the Amhara regional state (map 5.1) and is endowed with a rich groundwater potential. However, unreliable rainfall and surface resources have caused major droughts, which have constrained development in the area (Endalamaw, 2009). Most of the farmers are still dependent on rain fed agriculture and spate irrigation, although groundwater systems are present and developing. The groundwater development for irrigation  purposes in Kobo is centrally organised via the Kobo-Girana Valley Development Programme (KGVDP). This programme is responsible for the management and overseeing the development of groundwater systems for irrigation (further explained in chapter 6.1.4).

    In Aybe kebelle, a modern drip/sprinkler irrigation system was installed in 2003. This system was one of the first projects of the Kobo Girana Valley Development Programme (KGVDP), but is having trouble with functioning since the start. The chair of the water committee explained during the interview how some farmers are changing the system, to get more water to their land. In the case of Aybe, one part of the sprinkler is being removed[6] and the drips (small holes) in the pipe systems are enlarged. These actions change the overall water distribution in the whole systems. Farmers who are further away from the water pump itself receive less water due to the modifications of the other farmers (Interview 4). The water committee and the government try to explain how the system is

---

[6] On the picture, the chair of the committee is explaining how the farmers are changing the sprinklers.

working and the water they are receiving is enough for the crops[7]. However, after all these years, the farmers still find it difficult to understand that the water they receive is enough for their crops. The chair and irrigation expert of the KGVDP think this is caused by the fact that the water is not always visible, especially with the drip system (Interviews 3, 4 & 29).

Map 5.1: Amhara region indicating Kobo (blue square) and Kombolcha (red square) woreda (UNDP, 2013)

The main problem is Aybe kebelle is an issue about the irrigation system between the water committee and a group of users. There is a group of farmers that is resisting to use and pay for the system. For these modern irrigation systems, the operation costs are higher than for other systems (electricity is needed to pump the water from the ground and there are also more maintenance costs for drip/sprinkler systems). Therefore farmers have to work on their land while they are paying for the system, to earn the money back. During the rainy season, the system is only used to supplement in cases of short rainfall. The disagreement between the resisting farmers and the water committee has gone so far, that last December, a group of farmers broke some pipes and connections intentionally during the night. Since then, the system is not working at all. The chair does not know exactly who the resistant farmers are, therefore it is difficult to punish this group for their actions and talk with them to solve the issue (Interview 4). However, they are able to influence the functionality of the whole system for (Interviews 3 & 4).

In this case, the most important factor leading to the conflict situation are the social and political relations within the community. There are issues going on and power relations present in the village that are influencing the relations between the farmers in the irrigation cooperative. However,

---

[7] In Aybe kebelle, the plots irrigated by the drip system receive 5 mm water per day and the plots with sprinklers receive water once every 8 days.

due to the short time spent in this village, the deeper social relations were not identified. The irrigation expert of the KGVDP was also suggesting that other issues in the village are related to the issues with the irrigation system (Interviews 3 & 4). The water committee is not strong enough to manage the system. The members of the committee are already in power for a long time and did not succeed in their main tasks. The issue between the water committee and a group of farmers is causing the problem and neither the water committee itself or with help of the local government and the KGVDP are able to solve the issue at the moment. Distribution is a minor issue and is related to the fact that farmers are trying to lead more water to their fields by adapting parts of the system. This is however not the main cause of the conflict situation. The management situation can be categorised as intra-community conflict with influence on the functionality of the system.

  b.   Diversion Raya-Azebo; conflict between farmers and the local government.

Raya-Azebo woreda is located in the Raya valley, a lowland area in Tigray regional state (map 5.2). The Raya valley is one of the resource-endowed parts of Tigray with respect to groundwater, fertile land, livestock potential and climate conditions. There is water available for irrigation in a sustainable manner and this is needed to supplement rainfall and surface water resources, which cannot fulfil irrigation requirement due to seasonal variability and unreliability. However, the developments of groundwater irrigation systems have been slow. Compared to Alamata woreda and Kobo woreda, who share the same valley system, Raya-Azebo woreda is less developed with respect to irrigation and drinking water systems (Endalamaw, 2009).

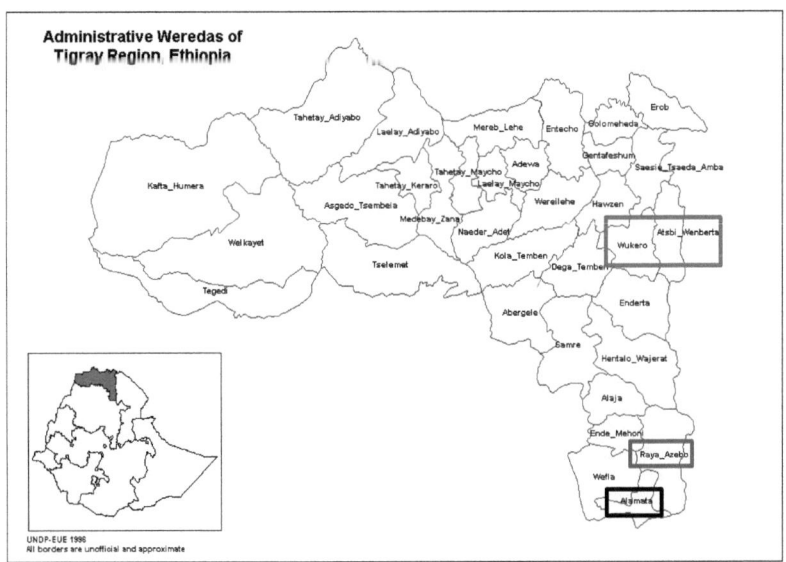

Map 5.2: Tigray region indicating Raya-Azebo, Alamata, Kilite-Awlalo (Wukero) and Atsbi-Womberta woreda (UNDP, 2013)

In Raya-Azebo woreda, a river diversion is supplying water for irrigation to two kebelles; in total about 800 farmers are depending on the diversion, which irrigates 125 hectares currently. One kebelle (Tsigea) is working on modernising the diversion system, by constructing a concrete canal system to organise the flow from the diversion to the agricultural fields. The farmers are providing labour and materials to contribute to the construction works. The constructing of the canal system is still going on, but one can already see how the water is flow in the canals is improved compared to the traditional hand dug canals. Tsigea kebelle is planning to expand the irrigated area within the kebelle from 80 to 120 hectares. Genete kebelle is the other kebelle using water from the diversion, but they do not have modern canals and are still using hand dug canals to transport water to the fields. In Genete, 45 hectares are now irrigated by the diversion. The two kebelles are sharing water source and are struggling about the water distribution. Both kebelles feel they have the right to receive more water than the other kebelle and Genete kebelle is even angrier now, because they see the improvement of the modern canals. According to the woreda office, Genete kebelle is not willing to provide labour and materials to construct the canals. The farmers from Genete say that they are now working on conservation activities and will work on the canals after the conservation works and it is not a case of unwillingness to work for the construction (Interviews 12, 13, 14 & 30).

In this case, there is no extreme scarcity of water according to the irrigation expert of the woreda agricultural bureau, although the water is not always enough to supply water to every farmer during the dry season. The issue is actually about the distribution of the water between the two kebelles and for Tsigea kebelle also the distribution within the kebelle. In the beginning of the diversion structure, there is a canal that has two branches, one leading to Genete kebelle and one leading to Tsigea kebelle. In this way, the water distribution can be regulated, by closing one of the two branches leading all the water available from the diversion to one of the kebelles. Both kebelles have their own traditional water committee responsible for the distribution within the kebelle. The distribution used to be even between the two kebelles, but the local government officials have changed this schedule when they started working on modernising the diversion system in favour of Tsigea kebelle resulting is complaints from Genete kebelle (Interviews 12, 13, 14 & 30).

The absence of an overarching water committee, or two registered water committees, can be seen as a lack of functioning of institutions. The water committees are not strong enough to handle the water management by themselves. Also the influence of the local government is contributing to that factor. There is also an issue in access to land, especially within Tsigea kebelle. There are some farmers who have larger part of land than other farmers. These larger farmers are mostly former army personnel (guerrilla fighters during the Ethiopia – Eritrea war) and are organised in smaller farmer unions. These farmers have land closer to the diversion itself. The small-scale farmers have land more downstream in the kebelle and are sometimes complaining about access to water. The traditional water committee in Tsigea can handle this issue, by sharing the water on an equal basis, despite the difference in land size. Also, the improvement of the canal structure will improve this situation (Interviews 12, 13, 14 & 30). Overall, the most important factors leading to the current situation of verbal conflict are distribution/access and functioning of institutions. The situation can be described as inter-community, upstream-downstream and community-government conflict.

These first two cases represent conflict situations. In the first case a group of farmers is involved in violation of the system and thereby limiting access to the system for other farmers. In the second case it is more about protests and a verbal conflict between farmers from the two kebelles and with the woreda government. Both are considered conflicts, since they are present in the day-to-day management. Sometimes there are cases where there are problems and disputes, but these are not as visible as in the first two cases. The following three cases describe situations of void, where sometimes elements of conflict of cooperation are present, but overall not much is happening. However, the current situation is not in favour of the system, because there are problems present locally with little attention which can lead to conflicts in the long term.

c. *Kugitelemlem + Selamgelasi; void situation with disagreements between the two kebelles.*

The woreda Alamata is part of the Raya valley and is located between Raya-Azebo woreda and the Kobo valley (map 5.2). The kebelles Kugitelemlem and Selamgelasi share a deep borehole pump with a traditional furrow irrigation system called Fascha; this pump got operational in January 2013. The water pumped by the deep borehole is transported via traditional hand dug canals and is irrigating 40 hectares currently. Farmers from the two kebelles use this deep borehole: 78 households in Kugitelemlem and 67 households in Selamgelasi.

The sharing of the water sources is resulting in some tensions and power issues between the two kebelles. According to the chair of the water committee, there is a complication in the contribution of the farmers from Selamgelasi kebelle; these farmers gave their contribution to their traditional leader. This traditional leader has no land within the irrigated area and is therefore not a member of the cooperative. As a result, he is not willing to give the money to the committee managing the irrigation system. Besides the money issue, the members of Selamgelasi are less active in the cooperative than the members of Kugitelemlem. Before the system was working, the members from Kugitelemlem were already actively involved in organising a cooperative. In a later phase it became clear that the scheme would also supply water to members of Selamgelasi kebelle. Also, in the committee Selamgelasi is less represented with only two of the twelve committee members are coming from this kebelle (Interviews 1, 15 & 16).

Besides the issues between farmers from the two kebelles using the water scheme, there are also other small issues. For the farmers, using an irrigation system means adjustments in their farming practice; where by rain-fed production the focus lays on production for their own households, in modern irrigation systems a change to cash crops is needed to pay for the use of the system and to improve the livelihoods of the farmers. During the visit to the Fascha system, it appeared that farmers did not follow the advice of the woreda agricultural bureau to produce cash crops. The chair of the committee explained that farmers here are not ready yet to change their production, but this may change in the future. Moreover, there is a discussion between the farmers and the agronomists of the woreda agricultural bureau; the agronomists recommend the use of fertilizer, since the soil is more intensively used due to the irrigation system (two-three times production in a year instead of one). The farmers on the other hand are convinced they do not need fertilizer for their land type and crops they produce. This situation can be considered as void, because there is a clear disagreement between the farmers and the agronomist, but there is nothing happening. Farmers are continuing with their own practices, while the agronomists continue with convince the farmers with their advice (Interviews 1, 15 & 16).

In the issue between the kebelles Kugitelemlem and Selamgelasi are mostly caused by the social relations between the farmers from both kebelles. There is no equal representation of both farmers groups and the farmers from Selamgelasi are in general less active within the cooperative. The system started working some months before the visit, therefore the water committee is still starting and not functioning as it should be yet. Also, the relation with the local government is in principal good, but they were late with determining the irrigated area of the system and have influenced the relation between the farmers of the two kebelles. There is also an issue of distribution, due to the inefficient hand dug canals, it can take a while before the water is reaching the outskirts and these farmers have to pay more than farmers close to the pump (the farmers also have to pay for the time it takes for the water to reach their land). The issue between the farmers and the woreda agronomist can be explained as a lack of awareness of the farmers on how to use the new system. This is expected to change when farmers are more familiar with the usage of the system. During the visit, the farmers worked on their first cropping season with use of the irrigation system. Once they see the possible benefit, they will probably change their choice of crops more in the direction of cash crop producing for the local markets. The situation can be defined as inter-community, with notice that the issue is happening on very small scale, concerning one system.

d. Berki watershed; a river shard by two woreda's. No real conflict, but rapid developments.

The Berki River originates from the highlands of Tigray. The communities in the catchment practice traditional irrigation for more than 100 years. The Berki catchment is about 410 km2 and is shared by two woreda's; Kilite-Awlalo and Atsbi-Womberta where the latter is located at the upper catchment and the former at the lower catchment (map 5.2). In both kebelles, there has been rapid development of irrigation structures since recent years and major plans for the future. The Atsbi-Womberta woreda has a plan to introduce about 100  motor pumps in the upper catchment area, with possible impact on irrigation projects in the lower catchment area. There are different diversions along the river. The river is also used for other purposes, like domestic purposes, productive purposes and environmental functions. There is a spring near to the Berki diversion that is being used by a church for spiritual purposes (holy water). The church capped the spring with the fear that it may be developed by the government to supply water for the Agula town (ECWP, 2007). The Berki watershed used to be covered by primary forest. Agricultural activities, fuel wood collection and free animal grazing caused deforestation and have had severe impacts on the ecosystem and hydrological conditions in the area (ECWP, 2007). There is limited communication between the different users and stakeholders and low awareness about the connectedness of the systems and the severity of the problems (Ibid).

The problems in the Berki watershed were extensively described in various documents for a pilot project of the Ethiopian Country Water Partnership (ECWP)[8] in the Berki watershed, especially around 2007. This project was addressing the issues in the watershed and promoting integrated water

---

[8] Country branch of the Global Water Partnership (GWP), promoting Integrated Water Resource Management.

resource management on local scale. During the visit in the Berki watershed, the project was already finished. The observation in the fields and the interviews with the farmers revealed no major conflict in the downstream woreda or between the upstream and downstream woreda. The river was flowing and green fields were present along the river[9]. There is currently no major issue of water shortage in the river. The difficulty in Atsbi-Womberta woreda is the mountainous landscape. In this highland kebelle it can be a challenge to transport the water from the rivers to the fields. In some kebelles in Atsbi-Womberta woreda, not all farmers are able to use water from the river for irrigation at the same time due to water shortages and difficulties in keeping the water in the woreda. From the observations in the field can be concluded that the situation has been improved in the Berki Watershed. The government officials gave training and this has helped to create awareness. The watershed conservation activities increased the water flow in the area. A lot of people interviewed confirmed that the flow of the river has increased and we could also see with our own eyes that the river had enough water, even in the dry season (Interviews 5, 17-21).

The problems were caused in the past due to scarcity of water in the river (explained mostly by introduction of new schemes and users and changes in the environment). The current issues are more related to distribution and access. In some parts along the river, there is still a problem of scarcity, more related to the capacity to store water with help of dams and diversions (economic scarcity). Due to watershed management, the water availability in the river has increased. Therefore scarcity is not a pressing issue anymore, but who has access and how water is distributed is still an issue in some parts along the river. There are no overarching rules to guide the access to the river for the different stakeholders. Moreover, the river is used by different uses and users. This is sometimes leading to issues between the different uses, with other words competition (multiple-use problem). Functioning of institutions is also influencing the situations. There are only a few water committees and most of them are traditional. Also the local government authorities are not paying an active role in controlling and overseeing the situation. The current situation can be seen as a void situation, because the management of the river is far from optimal and there is nothing happening to improve this situation. Everybody is continuing with the usage of water and the woreda's are even planning to expand the irrigation from the river. This situation is not desirable for the river and could possibly end in some conflicts when nothing changes. The Berki situation is a classic example of an upstream-downstream situation and is clearly inter-community since two woreda's are involved. This makes the case different from the other cases, where the problems are happening within the border of one woreda.

e.   Degem, void + solved conflict; still dissatisfaction and shortage of water.

Degem woreda is a highland woreda in Oromia Regional State (map 5.3). In Degem, most of the irrigation systems are using springs from the mountains and rivers. The area experiences two wet seasons, the small rain season is however erratic and unreliable. There is also a promising groundwater potential in the woreda, which is not extensively used at the moment. The springs are mostly used for domestic use as well as irrigation purposes. The multiple-use of springs and the sharing of springs between upstream and downstream kebelles are causing disputes and conflicts in the woreda.

---

[9] Visit was on 7-10 March 2013

Map 5.3: Northern part of Oromia region indicating Degem woreda (UNDP, 2013)

The village Tumano has access to a spring which originates in the nearby hills. The village was using this spring for domestic use and in the past, also for irrigation of fields close to the village. However, the spring is not sufficient anymore to cover both purposes. Therefore, the water committee decided that drinking water has the preference and irrigation is not allowed anymore, especially in the dry season when the water availability is low[10]. At the moment of the field visit (April 2013), the households were only allowed to irrigate some small area within their own compound. This was decided by the traditional water committee of five members. However, some visited households were also irrigating some land outside their compound and other households told us that they were not allowed to use water for irrigation at all. In Tumano kebelle 200 households are living, 120 households have their own connection in their compound. The other households collect water from three public water points in the village. The village is located on a hill, which is causing upstream – downstream issues within the village. It is difficult to store water for the upstream users; therefore a tank was constructed for night storage of water from the spring. Furthermore, there are rules and time agreements to divide the water between the different users in the village (interviews 22-24).

In the same woreda, the Tumano spring is also used for irrigation in Ano Kare kebelle. Part of the spring is going via a small river to the downstream kebelle Ano Kare and part of the spring is providing drinking water to Tumano village via pipes and tanks. Between the two kebelles, a conflict was present about the distribution of the water to Ano Kare kebelle. This conflict occurred around December-January. The water availability was not enough to support both kebelles at the same time. The usage of the water in Tumano kebelle disrupted the water flow to Ano Kare kebelle. The consequence was crop failure and that resulted in some fights. To find a distribution schedule, long discussions with involvement of the woreda government (Interviews 22-24).

---

[10] The preference for domestic water use before other uses is imposed by the government (MoWR, 2001).

This case presents a clear example of a scarcity problem leading to disputes and conflict between different users of the same source. This can be considered as economic scarcity, since there is potential to develop groundwater systems for supplementing the existing sources. The functioning of institutions is also contributing to the problems in the woreda. The existing water committees are traditional and have no well-developed rules and regulations to manage the water resources. It was for example not clear for all the villagers whether the private taps could be used for irrigation or not in Tumano village. Moreover, the woreda government lacks technical knowledge and financial capacity to develop more sources and change the scarcity situation in the area. In this case competition between different types of uses is present. This is an example of a multiple-use conflict. The small river shared between Tumano and Ano Kare is besides irrigation, also used for doing laundry and livestock watering. This situation has been a conflict before the distribution schedule between the two kebelles had been formulated. However, there is still some tension between different parties and economic water scarcity. Therefore, this case can be seen as a situation of void. The issues can be defined as an upstream-downstream conflict, intercommunity and multiple-use.

The three cases described above present different void situations, where various types of problems are present, but there is no outspoken situation of conflict or cooperation. The problems are not leading to negative influence on the systems at this moment, or the local government has tried to solve issues in the past, without creating a situation of cooperation or eliminating the sources of tensions. However, when the problems in the cases are not taken care of, the situations can jeopardise the sustainability of these systems. The last two cases describe situations were elements of cooperation are present, although in the past these cases also have experienced situations of conflicts and problems. These issues are however (almost) solved and the usage of the water system are going peaceful.

f. Diversion Kobo: solved conflict + new tensions between two kebelles

In Kobo woreda, several river diversions are constructed, providing water to larger groups of farmers during the year. These diversions are managed by the woreda agricultural bureau. One of the diversions in the Kobo valley is being repaired and expanded at the moment of the visit (March 2013). The diversion is currently irrigating 400 hectares in 08 Golina kebelle and started working in 2002. Some 80 hectares is irrigated by the night storage, the remaining 320 hectares is irrigated during the day, directly from the diversion. The traditional water  committee is responsible for making the distribution schedule for Golina kebelle. The farmers depending on the night storage receive water once every six days, whereas the farmers using the daily flow receive water once every four days. The canals are partly lined modern and partly traditional hand dug. Around 1700 households in Golina kebelle depend on the water from the diversion to irrigate their land. These households have small land holdings, but the irrigation opportunity meant a change in livelihood. Farmers told that they could built better homes, sent their children to school and pay for healthcare (Interview 25).

The recent expansion to the neighbouring kebelle 36 Mengello is causing for some disputes between the two kebelles regarding the water distribution. The new kebelle is claiming half of the water, although they are now only irrigating 30 hectares (they are planning to expand this area after the construction of the diversion has finished). This claim is not well received by the farmers and committee members in Golina kebelle and the woreda agricultural bureau had to intervene in the discussion between the two kebelles. Another major issue is the water distribution within Golina kebelle; there have been disputes between farmers about the water distribution. Farmers are cheating with the water distribution, by leading water to their own fields when it is not their turn. This is not only bad for the farmers who were supposed to get water, is it also bad for the functionality of the whole system. To change the water flow, farmers are sometimes destroying part of the diversion system. The committee has to repair the damages made by the farmers. Therefore, the committee has rules and regulations and fines for when farmers are caught (different amounts, between 50 and 500 birr[11]). In the past, these conflicts even resulted in sporadic physical fighting. There were some incidents before 2007, where in different incidents eight people got injured by knifes and one farmer was even killed in a fight (Interviews 3 & 25).

In this case, the major factor leading to the problems is the distribution of the water, between the two kebelles and within Golina kebelle itself. The diversion is irrigation a large area and with the maintenance of the diversion itself, plans are ready to expand the area in Mengello kebelle. The farmers in Golina kebelle are using the diversion already for more than ten years and know the benefits. The expansion is therefore seen as a threat, since they want to secure their access to water. At the moment, there is no scarcity of water to provide water to the farmers, but with the expansion there is a fear from the farmers that this might become an issue. The irrigation expert of the woreda agricultural bureau explained that this is not expected, since there have been studies to investigate whether the expansion is possible (Interview 29). This case study represents an inter-community situation, with some elements of upstream-downstream problems, although it is not as present as in other cases described in this section.

g. *Kombolcha; solved conflict + active attention for conflict resolution.*

Kombolcha woreda is located in the eastern part of Amhara regional state (map 5.1). The major water sources for irrigation and domestic use are springs and rivers. In general, the Amhara region is depending on surface water for irrigation, due to the presence of major rivers and the Tana Lake (Interview 9). The local NGO Water Action constructed a diversion in 09 kebelle, which is irrigating 40 hectares of land whereof 65 households depend on. Since the construction, the number of households has been expanded from 6 in 2010, to 46 in 2011 to 65 now. The diversion is focused on irrigation, although it is sometimes used for domestic use and livestock watering as well. The official drinking water point is far away for most households, so the temptation to use the water from the diversion for drinking is present. The use of the diversion for drinking water and livestock watering has not led to disputes so far. There is still enough water for irrigation for all the farmers. There is a water committee of twelve men responsible for the management of the diversion. The distribution of water is somewhat differently organised than in other irrigation systems. The irrigated area is divided in three zones of 22, 22 and 21 households. One member of the committee is responsible for the zone. Each zone gets water for three days during

---

[11] 1 Euro is 24.3399 Ethiopian Birr (XE currency convertor, 29-04-2013: http://www.xe.com/currencyconverter/convert/?Amount=1&From=EUR&To=ETB)

the day and night. In each zone, the water is again divided per day, for seven-, seven- and eight households. This distribution schedule is clear for all the households and the committee receives few complaints. In most other irrigation schemes, distribution is one of the main issues of complaints and disputes (Interviews 26-28).

Although the water distribution seems well organised, this system also experienced problems in the past. There was some conflict around the diversion related to the expansion of the irrigated area. Water Action wanted to increase the irrigated area by constructing a river crossing to irrigate some hectares on the other side of the river. For that part of the kebelle, it is difficult to use water directly from the river for irrigation, since the river itself is lower than the land. The farmers who used the diversion at that time were afraid that there would be not enough water for them and the people behind the river crossing. This fears were discussed with the farmers and in the end, the river crossing was constructed. At the moment of visiting (9 April 2013), the river crossing was destroyed by rain and also the diversion itself had got some damage from the rain. The farmers are now waiting on the repair, but it is not a major issue at the moment, since the small rain is present this year (Interviews 26-28). Therefore the crops are not lost or in danger of losing.

This case study represents the most positive case during the research. Although there were minor problems in the past, the way they are handled resulted in a peaceful situation at the moment where farmers can discuss their problems openly with the water committee and the water committee receives support from the woreda government when they need it. The farmers know when they will receive water and where to go for complaints. The main factors causing the past conflict were distribution and access the water, combined with lack of awareness of the farmers of the consequence of the river crossing. At this moment, there is no problem or conflict, except the technical problem which will be resulted soon according to the different stakeholders (Interviews 26 & 28). The current peaceful situation is also a result from the good social relations in the community. The community members are treating each other with respect and have open meetings about their problems. The past conflict can be defined as an upstream-downstream problem happening intra-community.

The last two cases represented more peaceful situation, where the local government is playing a positive role in dealing with tensions and problems, which were also present here. The major difference is the way the problems are dealt with and are transformed in a situation where farmers can discuss their problems with the water committee and the local government is supporting the water committee if needed. Case f (diversion in Kobo) also shows the presence of prior conflicts, which illustrates the dynamics of local situations.

This section described the different cases visited in Ethiopia from seven different woreda's and three different regional states. Table 5.4 presents the main characteristics of the described cases to create a clear and short overview of the main characteristics of the situations.

Table 5.4: Characteristics of the cases

Case name	A: Drip/sprinkler Aybe, Kobo	B: Diversion Raya-Azebo	C: Kugitelemlem and Selamgelasi, Fascha water point	D: Berki Watershed[12]	E: Degem, spring development	F: Diversion Kobo	G: Kombolcha
Type of system	Deep borehole with modern system	Diversion	Deep borehole with furrow system	Different, diversions, dams, motor pumps	Spring and piped system	Diversion	Diversion
Year of construction	2003	2006	2013	Varies	2010	2002	2010
Ha irrigated	40.6	80 (Tsigea) + 45 (Genete)	40	-	Unknown	400	40
Number of households using the system	153	500 + 300	145	- households along the river	200	1700	65
Members of water committee	7	2 committees of 5 men per kebelle	12	- Different committees	5	10	12
Contribution[13]	All costs divided per ha	Contribution of labour and materials	120 birr once and 20 birr per month	Most of the time no contribution	12 up to 100 birr per year (public tap or own tap)	40 birr per ha per year	4 birr per 0.25 ha per month
Water distribution	Drip 5 mm p/day, sprinkler interval of 8 days	14 days Tsigea water, 9 days Genete	Time based, 1-2 hours per ha	Depends on system	Partly private taps and three public taps	80 ha with night storage, rest once in 4 days water	In three zones, local distribution
Problems in key words	Motivation, functionality	US-DS[14], sharing 2 kebelles	Sharing 2 kebelles, staring issues	Sharing issues, US-DS, no formal management	Not enough water for irrigation, sharing issue	Water distribution, sharing 2 kebelles	Technical issues, extending issue

---

[12] The Berki watershed represents a river and not one system like in the other cases, therefore this table not always relevant for this case

[13] 1 Euro is 24.3399 Ethiopian Birr (XE currency convertor, 29-04-2013: http://www.xe.com/currencyconverter/convert/?Amount=1&From=EUR&To=ETB)

[14] US-DS = upstream-downstream

5.2 Synthesis

The above section described the major issues around irrigation system observed in the different cases. However, the discussed problems are not the only problems that are present around irrigation systems. There are also cases of technical problems, especially around modern irrigation systems; the farmers need to work hard and produce three times a year to be able to pay the costs for operation and maintenance. When farmers are not working on their fields, farmers can lose money instead of earn money. Moreover, lacking maintenance or the wrong kind of maintenance can also result in changes in the water distribution. Some farmers are cheating with the modern system by enlarging the drop holes are sprinkler holes to get more water. This is affecting the water availability for other farmers, which can lead to a dispute. It is sometimes difficult for farmers to believe that drip and sprinkler systems provide enough water to the crops. The water is less visible, especially for drip systems. Also in other systems, it is important to maintain the system well, or invest in improving the system. A last technical point is the long time it can take for systems to be repaired. Sometimes, spare parts have to come from Addis Ababa. Moreover, the right technicians need to be arranged and paid for (Interviews 4, 33, 34 & 35).

Textbox 1: Maintenance issues in Alamata
Repairs of water systems can take long times. In one case, it took eight months to repair a water point which was operational for only several months. In Selamgelasi, there is a water system called Biruhtesfa and this system broke down three months after construction. After that, it took eight months to repair the water pump. This caused different losses for the farmers; crops, seeds and income. The water committee reported the problem several times to the woreda water bureau. The officials expected that they could use a crane from Raya-Azebo woreda, but this was not the case. After that became clear, the woreda officials in Alamata reported to problem to the regional water bureau and technicians from that bureau repaired the pump in the end (Interviews 15 & 31).

Most interviewed water committees complained about the high costs for electricity, fertilizer, pesticides and improved seeds. The farmers are not willing to use fertilizers and pesticides for their crops; they feel it is not necessary and are not used to do this in their traditional farming practices. But woreda agronomists are encouraging the farmers to use the agricultural inputs and change to cash crop production. The irrigation structure is expecting a behavioural change from the farmers, which can take a while. The farmers will use the soil more intensively and need to adapt to the new water system. This problem was present in most of the deep boreholes visited during the fieldwork period. The response from the different woreda agricultural experts is that the agricultural inputs are needed to ensure a fertile soil and cash crops are needed to make money. Model farmers are helping in this development; farmers are more eager to learn from other farmers than to learn from woreda experts (Interviews 1, 4, 16 & 31-35).

Besides these problems, there are some positive developments. Watershed management is increasing the availability of water locally. In Kobo, this was mentioned, as well as in Kilite-Awlalo and Atsbi-Womberta. Also in Degem and Kombolcha, the government is investing in watershed management to improve the water availability in the area. Furthermore, the irrigation development means often a real improvement of the livelihoods of the rural farmers. Various farmers told they could increase their income with the irrigation opportunities. Because the farmers are able to produce crops two-three times a year and are less vulnerable for drought by using supplementary irrigation, they have now money to improve their houses, sent their children to school and pay for their households' health care. Therefore it is also important ensure sustainable and peaceful use of the systems (Interviews 12, 13, 20, 36 & 37).

Textbox 2: Positive example of Abreha wa Atsbha

Abreha wa Atsbha is a kebelle in Kilite-Awlalo woreda, known for its good practices in water management. Although the area is dry, the kebelle is able to use irrigation for production of crops during the dry season since 2003. This was not always the case, at some point farmers were moving to other areas to have a better livelihood. According to the village chairperson, the situation has improved and farmers are happy to live in the kebelle (Interview 36). There are a lot of hand dug wells (630) and also irrigation from the river is used. Initially, the hand dug wells were dug for individual use. The major success factor in this kebelle is the MERET project (Managing Environmental Resources to Enable Transition), a project to improve the local water situation. The kebelle was dry and had major problems with the water supply. But with the water conservation activities, the situation has improved. The kebelle is green now and currently 72% of the arable land is irrigated (492 hectares). The drinking water coverage is 78%. According to the village chairman, the MERET project was successful, because it gave freedom to farmers to conserve their own land and used a bottom-up/participatory approach of the involved NGOs. They did not come with a predefined plan, but discussed with the farmers what they wanted and took that into account (Interviews 17, 20 & 36).

When reflecting back on the described cases, it can be observed that the cases differ regarding socio-economic context, water resources and type of irrigation systems. However, most cases are in one way or another struggling with how to manage the systems and how to solve disputes and problems. In almost all cases, distribution and access to water is a factor leading to disputes and sometimes conflicts. Especially when diversions and other surface water systems are shared by two kebelles, disputes about the water distribution will arise. Every kebelle thinks they deserve more (or at least even) water than the other kebelle. Water resources do not respect administrative boundaries; therefore this factor will always be present. Also distribution within one system can be an issue. Sometimes, farmers complain about the water distribution or are even changing the system to transport more water to their land; this is often done by changing the traditional canals (which are open and close) or extending water distribution from drip and sprinkler systems. There is also a difference seen in how water committees are distributing the water between the farmers. The water committee in Kombolcha (case g) has a really clear scheme where the land of the farmers is divided in different zones and in each zone the land is again divided per household. In this way, the farmers know when they receive water. Other schedules are less transparent for the farmers. In some cases, farmers can request water in the morning and the person responsible for operation will make the schedule for that day. In other cases, the water committee makes the schedule. With modern drip/sprinkler systems, the area is divided in plots, which can be regulated. The clearness of the distribution schedule has influence on the number of local disputes; when farmers know when they will receive water, they will not complain as much or try to cheat.

The social and political relations within the community are also an important factor, not always directly causing the problem, but influencing in the dynamics of the issue. If there are some issues within the community, these can lead to problems in management of the system. On the other hand, strong social relations can be a social-control mechanism helping to solve conflicts and detect problems. In practice, this factor was only in two cases identified, one time positive (Kombolcha, case f) and one time negative (Aybe, case a). These two cases represent the two extremes with regard to conflict and cooperation. Probably, this factor will play a role in all the cases, but in these two extreme cases, the factor is more present and influencing than in the other cases.

Text box 3: Personal disputes in Alamata
During the interview with the chair of the water committee and other farmers in case a (Kugitelemlem versus Selamgelasi), one farmer came to us and told his dissatisfaction with this water committee. The farmer has land just outside the irrigated area and he is not getting water for irrigation. He thinks it is because the committee members do not like him and wants to take revenge on him. The farmer just got out of jail and the people in the committee knew the victim (possibly murder). He claimed that his former wife got water on the same land he is now using, but he does not get water now. The farmer told his version of the problem very heated and angry, while the committee members tried to get him away from us. The chair explained that the woreda is deciding on which land gets water and where the border is. So it is not the decision of the committee and they cannot change the decision of the woreda and the former wife did not get water on the land at all. This case of dispute is exceptional and not representative for the situation in north Ethiopia. But it is illustrating that not all conflicts and disputes are caused by factors directly related to water issues, also personal issues can be the reason for conflict (Interview 16).

In general, functioning of institutions is always a factor involved in the dynamics of problems. This is related to the factors of distribution, competition, economic scarcity and social/political relations within the community. The factors scarcity and competition are less visible in the empirical cases. Scarcity was in none of the major cases the cause for the conflict or tensions, although it played a role in the Degem case (in the form of economic scarcity). This is in line with the theoretical discussion, that scarcity can be one of the factors, but is often not the only factor causing conflicts (Chapter 3). Competition is more visible in situations where water is used for multiple purposes, like surface water sources used for domestic and productive uses (for example the Degem case). In the case of surface water systems, upstream-downstream issues are often present, as inter-community issues, since these systems are often shared between two or more kebelles. In the case of a small system, there is more often an intra-community problem. In some of the cases, conflict between the community and government is present, most clearly in the Raya-Azebo case (more extensively described in chapter 6). The relation between the cases and factors is visualised in figure 5.5.

The cases have been presented in the continuum from conflicts to cooperation. The first and the last case present extreme cases of real conflict and cooperation, but most conflicts are in the middle and present some kind of void; situations where elements of conflict and cooperation are sometimes present, but the situation is blurry and the water system is not management sustainably. Furthermore a difference between groundwater- and surface water systems can be observed. In the case of groundwater systems (case a, c and e), the issues are mostly related to distribution issues. This seems to be related with the efficiency of irrigation systems. Groundwater systems that use traditional furrow canals to distribute the water are not very efficient and it will take a long time for the water to reach the plots far away from the water pump and are difficult to monitor (case c and e). Other farmers can easily cheat and lead the water to their own lands. In water flow, you see a major difference between hand dug canals and concrete or plastic canals. The higher efficiency could increase the area of irrigation. If constructed well, these modern technologies can help to distribute the water more efficiently.

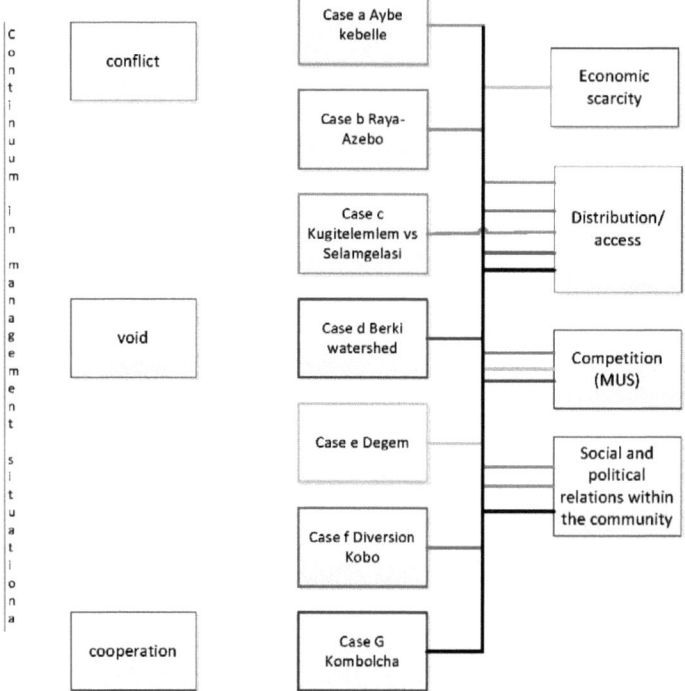

Figure 5.5: Framework of the cases and factors

The presented cases show that situations of conflict and cooperation and the intensity depend from place to place, influenced by the local context and different from time to time. Conflict is also related to the development of the farmer; if they adopt the technology, they will see the benefit and want more water (Interview 8). The farmers need to be aware about the benefits of the system and how to use it. The flowering time is a critical period in the cropping process, whereby farmers need access to water. During this time, the tensions and disputes between farmers can escalate more easily. Besides timing in crop seasons, the time in the year also makes a difference. Ethiopia has two rainy seasons, but the small rainy season is not always reliable. During and the first months after the main rainy season, the risk on conflict is low, because water is available, also via surface sources. In the period January-April, conflict may occur, especially when the small rainy season is small or absent (around March). This is related to the availability of water, or with other words, water scarcity. However, this is not directly observed from the cases itself (maybe caused by the fact the small rainy season was present during the visits). But it will have an influence, since problems can escalate or new problems can arise when the rain is not present in the right amount. It is also expected that this factor will become less appearing, when more farmers are depending on irrigation systems, which can supplement the rain. Last year the small rainy season was absent, but during the fieldwork this year, the rain started and the fields turned green. In Kombolcha, it was not even a big problem that the irrigation system was demolished by the rain. Due to the same rain, there is some time to repair the pump without losing crops (Interviews 9 & 10).

6. Role of local institutions

On different levels of the Ethiopian government, organisations are involved in the study, design, construction and management of irrigation systems. In this chapter, the roles and responsibilities of different institutions will be central. The first part will describe the roles and responsibilities of the different government organisations regarding irrigation development and management. The second part of the chapter will look at the role of the local government organisations in the different management cases. The analysis in this chapter is based on policy documents and interviews government officials from the federal, regional and woreda level in order to answer the second sub-question: *What is the role of local government institutions in the local management situations?* In this chapter, the focus is on the daily management of irrigation systems and especially the role of the local government in the different cases. The next chapter will go into the specific ways on how to solve conflicts, which is part of management. Conflict situations are considered as extreme situations where a special approach is needed, which falls outside the general management aspects of irrigation systems. On the other hand, managing water resources equals managing conflicting interests. For the discussion in this thesis, a distinction has been made between the daily management of irrigation systems (with the supporting role of the local government) and conflict management, which will be discussed in the next chapter.

6.1 Roles and responsibilities of the different government levels
In chapter 2, the national policies regarding water resource management have been discussed. These policies form a framework in which the different government levels operate to implement the plans and reach the policy goals. Below, their responsibilities will be discussed.

6.1.1 The Federal Government
On federal level, the Ministry of Water and Energy[15] is directing developments towards medium- and large scale irrigation systems[16]. The Ministry of Agriculture is focusing on support for small scale irrigation systems; the regional Water Bureaus are mostly responsible for the construction and design of small scale irrigation systems. The Ministry of Water and Energy is also responsible for formulating national water resource management policy, strategy and action plans (Arsano, 2010). Furthermore, the Ministry is monitoring the overall developments in the sector and providing technical support to mostly the Regional Water Bureaus. The Ministry of Water and Energy is using public or private contractors to execute their projects (Interview 11).

The Ministry of Agriculture is responsible for supporting the regional offices in agricultural activities. The main responsibilities regarding irrigation of this ministry are (FDRE, 2010, p.5641):
- Promote the expansion of extension and training services provided to farmers, pastoralists and private investors to improve the productivity of the agricultural sector.
- Formulate and facilitate the implementation of a strategy for natural resource protection and development through sustainable agricultural development.
- Expand small-scale irrigation schemes to enhance agricultural development.

The Ministry of Water and Energy is responsible for the promotion of the development of water resources and energy supply. In the field of irrigation, the Ministry is involved in the study, design

[15] The Ministry of Water and Energy is existing since 2010. Before 2010, the Ministry of Water Resources was responsible for water resource management and the Ministry of Mines and Energy for the energy component.
[16] Small-scale irrigation < 200 ha, medium-scale between 200 and 3.000, and large-scale > 3.000 (interview 6).

and construction phase of especially medium- and large scale irrigation schemes, whereas the Ministry of Agriculture is involved in small scale irrigation and the management of the various irrigation schemes. The main responsibilities of the Ministry of Water and Energy related to irrigation are (FDRE, 2010, p.5650):

- Undertake basin studies and determine the country's ground and surface water resource potential in terms of volume and quality, and facilitate the utilization of same.
- Carrying out of study, design and construction works to promote the expansion of medium and large irrigation dams.
- In cooperation with the appropriate organs, prescribe quality standards for waters to be used for various purposes.
- Support the expansion of potable water supply coverage; follow up and coordinate the implementation of projects financed by foreign assistance and loans.

6.1.2 Regional and Woreda level

The regional water- and agricultural bureaus are responsible for the irrigation developments in their region. The responsibilities of the water and agricultural bureaus are comparable with the ministries. The main task of the regional water bureau is the study, design and construction of water schemes and monitor small-, medium-, and large scale irrigation projects in the region. The bureau is also selecting the sites for new schemes, based on water resource availability, willingness of the local farmers, suitable land and livelihood characteristics. In some cases, the regional agricultural bureau is being consulted on selecting sites and irrigation systems. The actual construction of the water schemes is performed by public and private contractors. After construction of the water scheme, the regional agricultural bureau is responsible for making the irrigation system operational and overseeing the management of the system by the local water committee and the woreda agricultural bureau. The regional agricultural bureau is also involved in providing technical assistance and advising farmers about agricultural inputs and rehabilitating schemes. The regional bureaus oversee the activities of the woreda organisations and submit progress reports to the federal ministries, who then can track the developments in the country (Interviews 8 & 9).

On woreda level, there are also agricultural and water bureaus[17]. The woreda water bureaus are responsible for the study, design and implementation of small scale irrigation systems. After implementation, the systems are handed over to the community and the woreda agricultural bureau is responsible for overseeing the operation and management of the irrigation schemes (Interview 1). The woreda agricultural office is giving training to the members of the water committee on how to manage the systems and how to adapt their agricultural practices on the new systems (crop choice, agricultural input, etc.). If there is a technical issue, the woreda water bureau is again responsible for the repair. When the technical issue is beyond the capacity of the woreda water bureau, the regional water bureau is consulted for technical assistance. The woreda agricultural bureau is also involved in conflict resolution when the water committees are not able to solve the conflicts themselves. Sometimes, the woreda water bureau and woreda administration bureau are also involved in conflict resolution (Interview 9).

On kebelle level, there are Development Agents (DA's) responsible for direct assistance to the community. The DA's are part of the woreda government structure and communicate the

---

[17] The complete names are often Woreda Bureau for Agriculture and Rural Development and Woreda Bureau for Water and Energy, or Water Energy and Mines. But in the thesis, the names Woreda Agricultural Bureau and Woreda Water Bureau are used.

developments locally to the relevant woreda bureau. In each kebelle, there are on average three development agents working in the field of Natural Resource Management, Livestock and Crop Production. In general, all DA's working fields are related to water. Therefore, the DA's are all trained in irrigation management by the woreda and regional government (Interview 9). Most of the DA's are fresh graduates of the university. To summarise, table 6.1 is presenting the roles and responsibilities of the different levels of government in Ethiopia.

Table 6.1: Roles of the different governmental levels regarding the WASH and agricultural sectors (Butterworth et al., 2011, p.13).	
Level of government	**Roles and responsibilities**
Federal	• Formulation of policy, strategy, regulatory and planning mechanisms
	• Provision of approaches and support (preparation of guidelines, manuals, etc.)
	• Decision making on spending of national resources as well as allocating regional government budget from treasury
	• Coordination of the implementation of largest capital investment projects
	• Negotiating and signing of donors budget support and channelling to regions
Region	• Decision making on spending of regional resources e.g. allocations between woreda's (by regional cabinet)
	• Implementation of major projects and programs
	• Provision of technical support to zone and woreda's
	• Donor and NGO coordination
	• Follow up on progress of implementation of activities by woreda's (including those implemented with support from donors)
Woreda (district)	• Decision making on spending of woreda resources (by woreda cabinet)
	• Implementation of smaller projects
	• Following up the proper functioning of schemes and reporting to the higher levels e.g. where repairs are beyond capacity
	• Planning of different projects and coordination with donors
	• Supporting WASH committees technically and on scheme management and handling of collected monies
Kebelle (municipality)	• Planning and implementing
	• Support and follow up e.g. to WASH committees
	• Coordination: Kebelle managers, agricultural development agents (DAs), Health Extension Workers (HEWs), school directors and Kebelle chairperson works as *Kebelle* WASH team
Local (e.g. water user associations and WASH committees)	• Day-to-day management of schemes after 'handover'
	• Collecting fees
	• Small operation and maintenance

6.1.3 Local organisations

Amhara regional state has two main projects which are governed by their own project office; Koga project, where there are planning to irrigate 7.000 hectares via a river diversion and the Kobo-Girana Valley Programme (KGVDP), where the focus is on groundwater developments (Interview 9). The KGVDP is the other programme running in the Amhara regional state to develop irrigation systems. In the Kobo-Girana Valley, the focus is on the development of modern groundwater irrigation systems, since the valley has a high groundwater potential. Originally, the KGVDP was set-up in 1998 to support the development of groundwater-based irrigation system in the Kobo-Girana valley. In the beginning, the KGVDP was responsible for the design, construction and management of the groundwater irrigation systems. In 2009, the regional offices took over the responsibilities for the design and construction of the irrigation systems; the KGVDP is now responsible for the management and maintenance of the systems. Since the change in responsibilities, the focus also changed; from modern pressurized systems to furrow systems.

On local level, water committees are established to take care of the day-to-day management of the irrigation schemes. This committee is responsible for the operation of the system (some committees hire one or two operators to take care of that), the distribution of the water, collection of the money, maintenance of the system and resolution of disputes and conflicts (Interviews 1, 2 & 10). Most committees exist of six members and meet once in the month, or more often when there are pressing problems. The members of the cooperative pay most of the time a membership fee to enter the cooperative and monthly fees for using the irrigation system (Interview 9). The water committees get support from the woreda officials in how to distribute the water, which kind of seeds and agricultural inputs are needed and in case of technical problems, the officials give technical assistance. New water committees (especially of groundwater systems) should be registered at woreda level and have a constitution that describes how the system is managed and which rules and regulations are present at the system.

The traditional schemes, mostly spate irrigation systems for irrigating land during the rainy season, are generally managed by the farmers themselves. The village leaders may be involved in this, but the woreda officials are regularly not involved (Interview 6). The knowledge for constructing, maintaining and using these systems is also found locally, since people already use this way of irrigating land for long periods of time. One kebelle which is using spate irrigation has been visited (Tesoma in Alamata Woreda). In that kebelle, there are 14 spate irrigation sites. Each site is managed by two persons, who are called Abahagers (water fathers). These Abahagers are responsible distribute the water as equal as possible and coordinate the maintenance after flooding which happen regularly. Abahagers is also the term used for other traditional water committees. There are unregistered water user association who also elect 4-5 people for managing water systems. The Abahagers are the men in the committee responsible for the management and also for solving disputes and conflicts (Interview 1). In most woreda's, there is a plan to change the traditional Abagaher management to the establishment of modern registered water committees (Interviews 12 & 27).

6.2 Role of the government in different cases

In the different cases described in chapter 5, the local government had a role, varying from supporting and mediating in the conflict to influencing the conflict or even withdraw support. In this part, three cases with an influential or remarkable government role in the daily management are described; the withdrawing government in Aybe kebelle (case a), the influencing government in Raya-Azebo (case b) and the supporting government in the diversion in Kobo (case f). In the synthesis, the role of the government is compared to the other cases.

6.2.1 Aybe kebelle (case a)

The drip/sprinkler system in Aybe kebelle (see table 5.4) was one of the first projects of the Kobo-Girana Valley Development Programme (KGVDP) and is having trouble with the functionality of the system since the beginning due to actions of a group of resisting farmers. The water committee of seven members is responsible for the daily management of
the system and hired two operators to manage the water flow to the different irrigated areas (opening

and closing the water flow to the different fields). The chair has been re-elected for six times in a row and is losing patience in managing the system, he also feels threatened by the group of farmers who are demolishing the system. The chair explained that he lost the authority and control to manage the system. He is now not repairing the system, since he is afraid that some of the farmers will demolish it again. The chair has now the problem that the electricity company wants money for the electricity connection and the farmers are not willing to contribute money, since the system is not working. The chair was telling his side of the story with a lot of emotion; you could see he did not know what to do to change the situation to make the system working. He is willing to work on the land and use the system to improve his livelihood, but at this moment the situation is stuck and none of the farmers can use the irrigation system for their lands (Interviews 3, 4 & 29).

The local government (KGVDP office and agricultural bureau) were supporting the water committee in solving the technical problems. Officials from the KGVDP and the agricultural bureau have tried to solve the issue by talking to the farmers in Aybe kebelle to explain how the system works and what the benefit is for the farmers if they would use the system, without much success. The main problem is that the water committee and the government officials both do not know who exactly the group of resistant farmers are that demolish the system. The KGVDP is describing the place as difficult to manage and actually withdraw their support to the system. The agricultural bureau has withdrawn their support, since the farmers of the system are not complying with the rules of the agricultural bureau. The farmers are complaining about high costs for agricultural inputs which have to be paid on forehand. The agricultural bureau is not receiving as much complains from other system, which is why they have decided to withdraw their support from the system; they could use their time and resources more effectively in other systems. The water committee is not happy with the decision of the local government, but are also out of options to solve this issue at this moment (Interviews 3, 4 & 29).

According to the irrigation expert of the KGVDP, the group of farmers demolishing the system could have land in another irrigation scheme based on surface water, were the contribution is low compared to this system. Therefore, these farmers do not need to work on their lands to improve their livelihoods and they prefer to use it during the rainy season, to not have to contribute money for the modern system. This can be seen as an issue of land access; some farmers have only land in the modern system and need to work on it the whole year to have a secure livelihood, but are not able at the moment to do that (Interview 3). The irrigation expert of the KGVDP was willing to try again to solve this issue. That was what he said after the interview with the chair of the water committee in Aybe kebelle. How exactly he wants to do that, he did not know at the moment, but probably the interview with the chair gave him new insight in the situation (Interviews 3, 4 & 29).

6.2.2 Raya Azebo (case b)

In Raya-Azebo, two kebelles are using the river diversion with their own traditional water committees to manage the water flow in their kebelles (table 5.4). The division of the water between the two kebelles is however an issue. The local government has a prominent role in the current distribution schedule, with as result the dissatisfaction of some of the farmers. The dispute at stake is the distribution of water from the diversion to the two kebelles. There is a point in the beginning of the diversion (located in Tsigea kebelle), where branches can be open and closed to lead water to one or the other kebelle. The two traditional water committees were disagreeing about the division of the water between the two kebelles. Tsigea kebelle was claiming more water due to the fact they irrigate more area, but Genete wanted to remain the equal distribution schedule which was in place before the modernisation of the diversion. According to the woreda agricultural bureau, the water committees did not manage to solve this issue and the woreda officials intervened. At the moment, the two kebelles are in the process of developing cooperatives with elected water committees. Until there are new water committees, agricultural bureau made the distribution schedule by taking into account the water availability, the types of crops that are produced in both kebelles and the number of farmers using the system. The judgment of the officials was that Genete kebelle is producing more permanent crops and has fewer hectares under irrigation than Tsigea kebelle. Therefore, the water demand is not as high as in Tsigea kebelle. The division made by the officials is that Genete kebelle receives nine days all the water from the diversion and after these nine days, Tsigea kebelles receives water from the diversion for fourteen days. The distribution within the kebelle is still organised by the traditional water committees (Interviews 12-14 & 30).

The traditional water committee and farmers from Genete kebelle are not satisfied with the solution of the woreda agricultural bureau. The kebelle receives less water than Tsigea kebelle and the water committee has to go to the agricultural bureau to get a letter for the Tsigea committee to claim their right on nine days of water, since the division point is located in Tsigea kebelle. Until the committees and cooperatives are strong enough to handle this issue themselves, the agricultural bureau will enforce this distribution schedule and supervise the distribution of water. If there are registered cooperatives with committees for the management, the disputes between the two kebelles about the water distribution can hopefully be discussed by themselves. It will take some time to finish the process of creating registered committees. Until then, the farmers and water committee from Genete kebelle will keep on complaining to the agricultural bureau to show their dissatisfaction (Interviews 12-14 & 30). The distribution of the water in the two kebelles itself is arranged by the traditional water committees itself, the government is only intervening at the distribution between the two kebelles.

6.2.3 Diversion Kobo (case f)

As described in chapter 5, Kobo woreda has a large diversion, which is shared by two kebelles: Golina and Mengello (table 5.4). The diversion in Kobo is having problems regarding the water distribution in Golina kebelle, which is leading to disputes between farmers. Some farmers are cheating by leading water to their own fields when it is not their turn leading between disputes between farmers. The committee has to repair the damages made by the cheating farmers. The committee has rules and regulations and fines for when farmers are caught (different amounts, between 50 and 500 birr[18]). In the past, these conflicts and disputes even resulted in sporadic physical fighting. There were some incidents before 2007, where in different incidents eight people got injured by knifes and one farmer was even killed in a fight. The farmer who died was a father of seven children who had a dispute with a neighbouring farmer. The killer blocked the water to lead it to his own land; the victim was upset about that and this resulted in a deadly fight. The killer got punished by eleven years of prison via a process in court (Interviews 3 & 25). Furthermore, the local government was consulted to discuss how to move forward, what rules to change to make sure this is not happening again. After this incident, the committee enforced strict rules to prevent these incidents. Since 2007, it is forbidden to have weapons on the farmlands. There are guards responsible for checking the farmers on weapons. The actions taken have stopped the physical violence in the

---

[18] 1 Euro is 24.3399 Ethiopian Birr (XE currency convertor, 29-04-2013: http://www.xe.com/currencyconverter/convert/?Amount=1&From=EUR&To=ETB)

diversion area, but the questions remains whether banning weapons is a lasting solution for the violence. So far it stopped the violence, but maybe the root cause has not been sufficiently addressed. Farmers are still cheating with the distribution schedule sometimes and this is still leading to small verbal disputes (Interviews 3 & 25).

Besides the past conflicts within Golina kebelle, there are currently issues around the distribution of water between Golina and Mengello kebelle. This diversion is expanded at the moment and the plan is to extend the irrigated area in Mengello kebelle. Although the irrigated area is small compared to Golina kebelle (30 versus 400 hectares), Mengello kebelle is claiming halve of the water from the diversion. The two kebelles have their own traditional water committees for distribution of the water within the kebelle. These two committees discuss with each other the water distribution between the two kebelles. A year ago, they decided the following distribution schedule: three days water for Golina kebelle and one day water for Mengello kebelle. After this agreement, farmers from Golina kebelle chose a new water committee. The new committee was not satisfied with the distribution schedule of the old committee. Therefore, they changed the schedule to four days for Golina and one day for Mengello. However, both kebelles are still not satisfied about the current distribution. As a result, they are complaining to the officials of the agricultural bureau. The woreda will mediate by bringing both groups together in a meeting. The woreda officials also studied the water availability around the diversion and concluded that both kebelles can use the water from the diversion at the same time. This solution is however not well received by the two kebelles and therefore the woreda let this solution go (Interview 25).

Although there are problems and disputes around the diversion in Kobo, the local government is playing a positive role in the problems. There is a solution, but one of the parties not completely satisfied (farmers from Mengello kebelle). However, the intervention made sure there is no actual conflict now and there will be new discussions between the different stakeholders to work on these issues. The members of the water committees respond positive to the current solution and the commitment of the woreda agricultural bureau to discuss the issue more in the coming period. At the moment of the visit (March 2013), the farmers were using the system peacefully and waiting for the next discussions to come. The woreda agricultural bureau is also thinking about changing the current organisational structure around the diversion. There is an idea to develop one committee/ organisation to be responsible for the management of the diversion. In this way, the management responsibility will lay closer to the community and it will be easier to discuss issues between the two kebelles. This new organisation should exist of members from the two kebelles and will receive training and support from the agricultural bureau (Interviews 3 & 25).

6.3 Synthesis

The three examples discussed show different roles of especially the woreda officials in the management situations. The role of the local government in the distribution issue around the Kobo diversion represents a positive example. The woreda officials are listening to the different actors and the environmental aspects are also studied and taken into account in discussion between the two kebelles. Also Kombolcha (case g) represents a positive case with regard to the role of the government. Here the government played a positive role, in supporting the water committee in the management of the diversion. The agricultural officials acknowledge the risk of conflicts around water systems. When there are complaints or notifications of disputes, experts are going to the field to assess the issues on the ground and discuss the issues with the stakeholders (Interviews 26-28).

Sometimes the local government officials have a more passive role and are not directly responding to disputes around water systems. The modern irrigation system used by the kebelles Kugitelemlem and Selamgelasi (case c) was only operational for a few months at the moment of visit. The water committee already informed the woreda officials about the starting issues and tensions between the farmers of the two kebelles. Moreover, in this case, the decisions of the woreda officials could have influenced the situation and maybe even caused the current tensions between the kebelles indirectly. The woreda officials are responsible for determining the area which can be irrigated by the system. The farmers from Kugitelemlem were aware that the system was getting operational and were already starting with establishing the cooperative. When the system became operational, they discovered that also farmers from Selamgelasi would use the system. But at that stage, the farmers from Kugitelemlem were already working on establishing the cooperative and the farmers of Selamgelasi were only involved in a late stage. This could have caused the low participation of the farmers from Selamgelasi and the payment issues within the cooperative (Interviews 1, 15 & 16).

Related to the Kugitelemlem and Selamgelasi case, in the Berki watershed the influence of the government is not really present (case d). They supervise the developments in the woreda's and construct new diversions and dams if there is money, but there is less support and influence than in the other cases. The support of the local government in the daily management of irrigation systems is here less present than in other cases. There is no active discussion between the woreda officials in the two kebelles to make agreements on the shared water use. Moreover, there are little registered water committees to manage the different irrigation systems along the rivers. Along the rivers, farmers are using water from the river for irrigation without much guidance or management (Interviews 17-21).

In the case Degem (case e) the local government had influence in solving the local issues and disputes, comparable to the Raya-Azebo diversion where the government made the distribution schedule. The influence of the local government in the distribution schedule is in Degem resulting in resistance from the farmers from Ano Kare. Another issue is that the local government is not well aware about the possibilities in the woreda with regard to groundwater development. This is resulting in disputes and problems around existing systems, whereas this maybe is not needed since there are other sources available to supplement current water sources.

The discussion of the cases shows that the local government can influence the management situations in different ways. In some cases, the woreda officials had a positive role in the situation, by actively bringing the different parties together, looking at the demand and supply of water and discussing a solution for the situation; in most of the cases a new distribution schedule to guide the daily use of the system. In this way, there is a greater chance of acceptance for the solution. In cases where the woreda officials enforced a solution or distribution schedule, there were always people unhappy after this event. This was seen in the case of the diversion in Raya-Azebo (case b) and the shared spring in Degem (case e). Aybe kebelle (case a) represents again an extreme case, where the woreda officials are withdrawing their support for the irrigation system. This will not solve the problems in Aybe kebelle and is maybe the end for the system.

The water committees are the first organisation responsible for the daily management of the systems and can ask help from the woreda level if they have problems with the management. The water committees have most of the time a constitution describing the rules and regulations of using the water system. In the case of traditional water committees, the rules and regulations are often verbal, instead of a writing constitution. In Amhara regional state, the woreda agricultural bureaus have a cooperation department to support the water committees and help them with formulating a constitution (which needs to be approved by the whole cooperative). The cooperation department provides a basic constitution, which the water committees can adapt to their system. Also in Tigray and Oromia regional state, the farmers get support in the establishment of cooperatives and a constitution, but bot from a separate department. If farmers break the rules of these constitutions, monetary fines are used to punish that behaviour. Most of the systems also have one or two guards who are responsible for protecting the system and overseeing the distribution of the water. They are paid by the monthly contribution farmers are paying to use the systems. These guards are most of the time present for groundwater systems.

Besides the roles of the local government directly in the cases, there are also other trends which can be observed from interviews with the different government officials. There is a lack of cooperation and collaboration between different levels or organisations of the government. In Tigray and Amhara regional state, there are a huge number of constructed deep boreholes which are not yet in operation for a few years. The local farmers are seeing these boreholes every day, but cannot use then for irrigation. Some woreda officials explain that the lack of cooperation between different bodies of the government is causing that (Interviews 1, 15 & 29). Also, at local level the woreda water bureau and woreda agricultural bureau need to work together on irrigation issues. One of the difficulties is that different governmental organisations are involved in the study, design, construction and operation and management of irrigation systems. This requires cooperation and coordination between the different organisations. However, the different organisations are not working together as they should be. Therefore, it can take a while for systems to be repaired if there is a technical problem, or for systems to become operational after the wells have been drilled.

On woreda level, the bureaus are struggling to execute their job. Capacity and officials with technical experience are missing in some woreda's. Students who come from the university have a lot of technical knowledge, but lack practical experience. And most of the time they start working on kebelle and woreda level (Interview 7). Another major issue is the lack of budget for constructing new irrigation schemes. A lot of governmental institutions are depending on financial assistance from NGOs and other organisations. Furthermore, most of the water-related conflicts are very local and not documented at all. How higher the governmental level, how lower the awareness and knowledge about local conflicts and issues. And because the conflicts are not documented, knowledge about older conflicts will be lost when people leave the offices. Sometimes they officials heard about previous conflicts, but do not know the details.

When looking back at the definitions from the theory chapter, this chapter has shown the importance of functioning of different types of institutions. The federal government is developing strategies for the water sector and the regional government is trying to implement these strategies in their region. The woreda government is the most important government organisation involved in the dynamics of the management situations. On local level, cooperatives develop constitutions consisting of rules and regulations on how to use the system. These rules are important in the management of the system and are part of the local rules and regulations. The water committees can be seen as local institutions. The local rules the water committees make are important for guiding the daily usage of the systems. In conclusion can be said that different types of institutions are important to ensure a sustainable management of irrigation systems and all deserve attention in the design of new systems to ensure this sustainable management.

Textbox 4: The strength of local institutions; the Borena case

Ethiopia has many different tribes, some have strong cultural traditions and organisations to manage their own community. An example of a strong traditional community is the Boran tribe in Oromia region. The Boran tribe is known for its alternative natural resource based conflict management (Costantinos, 1998). The Borena woreda is a dry lowland woreda in the south of Ethiopia, where most people are pastoralists. This region has experienced major conflicts over natural resources, these conflicts are often most intense in periods of severe droughts. Most of the conflicts are about grazing lands and water availability. However, compared to the other large pastoral communities (Afar and Somali), the Borena society is relatively stable (Ibid). The Borena society has developed a well organised system of livestock, grazing and water resource management, which is helping to control environmental degradation (Ibid).

The Gadaa system is a social organisation based on age-grade classes of the male population (Edossa et al., 2007). The different age classes are related to different roles and responsibilities of men in society. Men between 41 and 48 are the elders that can be elected as Gadaa officials. The Gadaa system is responsible for guiding the social, political, economic and religious life as well as contributing to conflict resolution among individuals and communities. The social mechanisms of the Borena society maintain peace and order locally and manage the communal pastoral resource collectively (Costantinos, 1998). There are rules for use of the water resources, based on the season (dry or wet). The grazing lands and water resources are managed as common property, while the livestock is privately owned. There are four sources of water: flood water, natural pond water, and shallow or deep wells dug by hand. There are nine deep wells, which are the sources during the dry season, when the other sources are dried out (Ibid). The deep wells are supervised by a trustee and managed by a manager. The Gadaa has a unique position in tackling the interlinked problems of environment, welfare and conflict (Edossa et al., 2007). Disputes and conflicts are solved through mediation or arbitration by a council of elders. In this way, the Boran tribe found a way to manage the common pool resources in their society. However, the Gadaa system is a male-oriented, socio-political and cultural system; women are excluded. This system falls outside the formal governmental structure. There is only loose collaboration sometimes. It would be good if the state recognises and support such customary institution, since they have a long history in natural resource management and conflict resolution. Moreover, they have strong foundations in the communities (Ibid). When the government can work together with the traditional institutions in this case to develop the Boran land, this could mean a real improvement for the people.

7. Conflict resolution

Situations of conflict are not desirable for farmers using irrigation systems, but they sometimes can trigger change in the way water systems are managed. In some situations, disputes or tensions around water systems are not solved and is there a situation of void, which is in the long term not positive for the functionality of irrigation systems. All management situations are dynamic and can change from nature. Situations of cooperation can have known conflicts and also in void situation, conflicts may have arisen in the past. There are different ways how to solve these issues. To answer the third research question (*What kind of conflict management tools can be used to solve local conflicts?*), this chapter will focus on the mechanisms of conflict resolution that have been used in the different cases. The focus will lay on what happened, how the solution was received by the different stakeholders, and what the result was. This chapter will only look at conflict management experiences, other management issues were discussed in chapter 6.

7.1 Experiences from the cases

This part describes the experience from conflict resolution in the described cases. The cases of the Berki watershed, Degem and Kombolcha have the main focus, although the experiences from the other cases are also shortly mentioned in the synthesis. Not in all cases, conflict management has been present clearly. The three cases are chosen for their usage of conflict management, and represent different ways of dealing with conflict. In the Berki watershed (case d, table 5.4), NGOs and regional government organisations have tried to solve previous issue in the watershed with an extensive project. In this chapter, the results of the project on the watershed will be discussed. In the Degem case (e) the government and water committee have tried to solve the problems, but the question is whether these solutions have been developed in the right way and have the desired outcome. Lastly, the Kombolcha case (case g) is described as positive case where a local NGO intervened in the way issues are handled. From all these cases, lessons can be learned for conflict resolution in these local management situations.

7.1.1 Berki watershed (case d)

In the Berki watershed (table 5.4), water from the river is used for irrigation and domestic use for a long time (ECWP, 2007). The increased water use of the upstream kebelle Atsbi-Womberta and the different types of uses of the river have led to the presence of conflicting interests and potential conflicts between upstream and downstream users and between the different uses of the water. These developments made the Berki watershed an interesting pilot project for the Ethiopian Country Water Partnership to promote Integrated Water Resource Management (IWRM). The project started in December 2005 with the goal to develop the Berki watershed sustainably and solve the tensions along the river (ECWP, 2007; Interviews 20 & 38). The project wanted to address these issues and focus on how to sustainably use and manage water resources for all interest groups (ECWP, 2007). To do this, multi-stakeholder forums were organised at different levels (regional, catchment, woreda and kebelle levels). The goal was to raise awareness of all stakeholders, promote their participation and to develop an integrated development and management plan for the catchment. It was also the objective to regulate the water use more in the Berki watershed, since there was a lot of illegal abstraction of water (ECWP, 2007). For the project, different meetings were organised to bring the different stakeholders together. Also, baseline studies were conducted to assess the current situation and this was the basis for the establishment of the Berki watershed development plan. The mission of the plan is to protect the catchment, ensure availability of good quality water and promote sustainable management of the environment (FDRE, ND).

An example of an conflicting issues dealt with in the pilot project in Berki was the issue between the Church in Agula and the woreda water bureau in Kilite-Awlalo wanting to use the water for supplying water to Agula town. The spring near Agula town has a big discharge of 30 litres per second and only a small part of that discharge is needed to supply water to Agula town (around five litres per second for three communal wells). This spring was used by a Church to function as holy water for religious purposes. One priest of the church was not agreeing with the decision of the woreda water bureau to use the water for Agula town. He claims that the spring is holy and belongs to the church. The priest used his power to oppose the decision of the woreda water bureau. Officials from the regional water bureau came to Agula and Wukro (woreda capital) to talk to the different stakeholders and see what kind of solution was possible. In the end, one priest of the Church was against the sharing of the holy water and was using his power to influence the people in Agula to fight for him. The woreda and regional water bureaus decided to look for another source and leave the holy spring to the church. According to the woreda water bureau, the priest was not open for the arguments of the experts. Now, a new source has been found for Agula town and is under construction. In the meantime, people from the town are using water from the river for domestic purposes, using water tanks to transport the water from the river to the town (Interview 20, ECWP 2007).

At the moment of the fieldwork (March 2013), the project was finished. The observation in the fields and the interviews with the farmers revealed no major conflicts in the area. One official explained that the situation of conflict which was described in the project documents was caused by the rapid irrigation developments in the upstream woreda. That had a major impact on the water availability. Due to watershed conservation activities, the water availability is increasing, which is reducing the risks of conflicts (Interviews 17-21). Most of the farmers use motorised pumps to pump the water to their fields or are depending on diversions/dams. In the upstream kebelle, the major challenge is to bring the water from the river to the fields. In both woreda's, the farmers are not organised in cooperatives, there traditional water fathers (Abahagers) are often responsible for the management. In one part in the upstream kebelle, not all farmers can use the water in the same season to irrigate their lands. But the farmers accept this fact and follow the rules of the water fathers. There is also an issue of small land holdings and especially young people have to rent land from older farmers to be able to produce crops (Interviews 5 & 21).

From the observations in the field can be concluded that the situation seems to be improved in the Berki Watershed. The watershed conservation activities increased the water flow in the area. The interviewed farmers and officials confirmed that the flow of the river has increased and we could also see with our own eyes that the river had enough water. However, the two woreda's are still sharing one river without much discussion and rules about the water use. There are no formal agreements on the water distribution between the two woreda's. This void of no clear rules is maybe not the best way of managing the river, where a lot of people depend on. On the other hand, the real conflicts and disagreements between different stakeholders and different uses seem to be solved or disappeared, mostly by talking and by the increasing water availability in the river. It seems nothing has changed much since the project of the ECWP, although the plan was to develop a water authority and bring the various stakeholders together and develop the watershed in a sustainable way. The University of Mekelle is now in the process of evaluating the impact of the project (Interview 41).

Regarding conflict resolution, multi-stakeholder platforms were used. Moreover, mediation from the local government was sometimes needed, for example in the case of the church in Agula town. The two woreda's are however not involved in conflict prevention. And also the traditional water committees are still not really aware about the importance of conflict prevention and resolution.

7.1.2 Degem (case e)

In Degem woreda there are disputes between villagers in Tumano kebelle and upstream-downstream issues between Tumano and Ano Kare kebelle (table 5.4). These two issues use ultimately the same source (a spring in the hills of Tumano Kebelle), but are approached differently in the way the issues are dealt with. The water distribution in Tumano kebelle is managed by a water committee of five members (elderly men of the kebelle). This committee is responsible for the distribution of the water to the 200 households in the village; for the three communal water points and the 120 private connections in the village. The village is located on a hill; therefore there is an upstream-downstream issue within the village itself. The committee also decided to stop irrigation from the spring supply to the village, since there is not enough water for domestic use and irrigation. There are also plans to install water metres, to control the water usage of the households with a private connection, but this solution is too expensive at the moment (Interviews 22 & 23). In this issue, the local water committee was capable of handling the issue without woreda level interference.

In the other issue, the distribution of water between Tumano and Ano Kare kebelle, the woreda officials had to intervene. The irrigation practices in the two kebelles is hardly organised, legal water committees are missing and only Ano Kare has a traditional water committee which is concentrating on managing the irrigation practices. Therefore, the local communities were not able to solve the dispute about the water distribution. Especially farmers from Ano Kare were complaining about not receiving enough water for their irrigation practices. The woreda level talked to the different stakeholders (the farmers from both kebelles and the water committee in Ano Kare) and made a distribution schedule. The upstream kebelle (Tumano) receives water for five days and night and the downstream kebelle (Ano Kare) for two days and nights in one week. The days that Ano Kare gets water are market days for Tumano, so the farmers are not around anyway. This became clear when Tumano kebelle was visited on Monday; there were no farmers on the fields. The farmers from Ano Kare kebelle are not happy with receiving less water than Tumano kebelle, but have accepted the solution and stopped with their actions and complaints for now (Interviews 22, 23, 24 & 40).

The issues in Degem woreda are calmed and distribution arrangement has been made to organise the use of water. However, in Tumano kebelle, the water from the spring cannot fulfil the current needs of the farmers and households. The villagers are not irrigating their lands at the moment and this is leading to a loss in income. Moreover, the demand of the farmers in the two kebelles is likely to grow, since more and more farmers see the benefit of irrigation and want to expand the use of it. On the positive side, the woreda is endowed with good groundwater potential, which is easily to develop by constructing hand dug wells and other types of shallow systems. There are some farmers who are trying to construct their own hand dug well in Tumano kebelle, to be able to irrigate their lands all year long. These developments are slow and the farmers receive little to no support or technical assistance from the local government. The officials from the woreda water bureau have a lack of knowledge about this potential and are not actively trying to develop the groundwater potential. The officials spoken to did not know about the attempts to construct the hand dug wells

(Interviews 23 & 40). Therefore, it can be concluded that part of the conflicts here are caused by the lack of knowledge and awareness of the local government to develop more water sources for domestic uses and irrigation (economic scarcity). In this way, existing conflicts could be resolved and new tensions prevented, although the new sources can also mean a new risk of tensions.

7.1.3 Kombolcha (case g)

The local NGO Water Action constructed the river diversion in 09 kebelle in Kombolcha woreda (table 5.4). Together with the irrigation department of the woreda agricultural office, Water Action paid special attention to conflict resolution and prevention in this project. The irrigated area of the diversion has been expected several times, resulting in some tensions between farmers on access to the water. To solve this issue, Water Action and woreda officials came to the field to explain to the farmers that there was enough water to make a fair distribution mechanism. Their conclusion was based upon feasibility studies of the water source. The different stakeholders were brought together using traditional mechanisms of community meetings and discussions. After this intervention, the conflict was solved and the river crossing was constructed (Interviews 26-28 & 39).

This conflict situation and resolution is an example illustrating the approach of Water Action and the woreda agricultural bureau. Both parties acknowledge that there are always conflicts around water systems, especially in times of low rainfall. Last year, the small rainy season was absent in this region of Ethiopia, causing more local conflicts in the woreda. This year, the small rainy season is present, resulting in less conflicts. The diversion in Kombolcha has currently technical problems caused by heavy rainfall; the rains destroyed the river crossing and some minor problems at the diversion itself. But these issues are not acute problems that will lead to crop losses, since the rain is presents and takes care of the necessary irrigation at the moment. The woreda agricultural office is planning to solve the issues with the diversion fast, although the repair of the river crossing can take some time. The change in practice and attitude of the farmers could result in more competition and possible conflicts in the future. Farmers are adopting the irrigation technologies; this will lead to more water use and possibly more pressure on water systems (Interviews 26-28).

To deal with the competition and the conflicts, Water Action is focusing on conflict resolution. In first instance, local water committees and village leaders should try to solve the problems by bringing the different stakeholders together and discuss the issue. Development agents and local leaders can perform a role of mediator and oversee the problem. If the problem is too complicated or beyond the capacity of local structures, the woreda level will be asked for help to mediate in the problem. Local leaders of water committees can write a letter to the agricultural bureau to inform them about the problem. If it is needed, the woreda officials will travel to the field to observe the problem and talk to the different stakeholders in the conflict. The local leaders and development agents are also asked for their opinions and observations. If it is needed, the involved parties are gathered by using local systems of community meetings to discuss the issue all together. In that way, they can propose solutions and solve the issues at stake. The focus on local practices first is believed to reduce the number of major conflicts. In first instance, the water committees are responsible for the conflict resolution. However, if an issue is too big, for example between two kebelles, the woreda level will intervene (Interviews 26-28).

Within the water cooperative, there is attention for resolving conflicts early. Every month, the whole cooperative of farmers is gathering to discuss problems. Every two weeks, the committee is discussing management, maintenance and operation of the diversion with the Development Agents and one local staff member of Water Action. Sometimes after a meeting, a field day is organised to give training to the committee/ farmers or give attention to a problem. In this way, the committee and the local government try to prevent conflicts and improve the performances of the farmers. The approach of the local NGO and the woreda agricultural office respect and use the local gathering mechanisms and leaders in solving conflicts and tensions. This fact in combination with their acknowledgement of conflicts and their analysis of the influence of rainfall makes this woreda and project an example for other woreda's. Although there are also disputes here, there are mechanisms to detect them early and solve them in a way that the local community will accept. The water committee also has the respect from the other farmers and works closely together with the local leaders and the woreda government.

7.2 Perspective of government organisations

During the interviews with government officials on different levels, there were questions directed towards the occurrence of conflicts and the role of the officials in solving these conflicts (appendix I). It appeared that the higher levels of government are not aware about the number and nature of conflicts happening locally. Only major conflicts they have heard of, but the fact that there are a lot of local conflicts (or situations of void) going on locally, was not known (Interviews 6-9 & 38). The lower levels of government are more aware about the local conflicts. Unfortunately, documentation about these documents is barely present. As a result, the conflicts happening a few years ago are forgotten when experts leave the area and there is hardly room for learning lessons from these situations.

Most of the regional and woreda water bureaus do not give training on conflict management. There is a lack of clear water laws and water user rights, but this is very important regarding the occurrence of conflicts. The water committees have the most important role for conflict management. Only when the conflict is arising on a larger scale, for example if two woreda's are involved, the regional level can intervene in conflicts. An engineer of the Amhara regional agricultural bureau only knows three concrete conflicts, but he is aware that there are a lot more conflicts occurring in the region (Interview 9). On federal level, the Ministry officials only knew about medium and large scale conflict, but not about the local issues. The Ministry of Water and Energy is mostly involved with designing and constructing irrigation systems and is not involved in the management of systems or conflicts which arise after construction. During construction there are often issues about land ownership and compensation for land expropriation (especially present in medium and large scale projects where investors develop a private business). Especially around major projects, the local population is protesting. They are closely following up the developments and try to find solutions (Interview 6). But these kinds of conflicts are different from the local conflicts around small-scale water schemes. These local conflicts barely reach the level of the Federal government, they are solved on lower levels of government.

The Amhara Agricultural Bureau has a manual for participatory development and management of small-scale irrigation systems for the Koga project[19] (Verheijen, 2011). Conflict resolution and prevention are mentioned in this manual a few times. The key message is that better management is leading to more efficient operation of the systems which is leading to less water-related conflicts. Moreover, water related conflicts would be solve easier and quicker (Verheijen, 2011). This manual is only used for the Koga project and in the cases of this research, not attention has been paid for conflict

---

[19] One of the two main projects of Amhara regional state, besides the Kobo-Girana Valley Development Programme.

resolution and prevention during the establishment and training of the cooperatives. This could be something to be improved, since it is essential for successful management.

On woreda level, the awareness is larger, but the officials are not always willing to talk about the number of conflicts or their role. They have after the local water committees the greatest role in conflict resolution and know best what is going on locally. The Kombolcha agricultural bureau is taking conflicts seriously and was directly admitting the incidence of conflict. This is leading to a more proactive and efficient approach to deal with conflicts. On the other hand, there are woreda's with a more passive approach, which are also aware about conflicts happening in their woreda, but are less aware about the details and are less actively involved in solving the issues. An example is Degem woreda, where the woreda water bureau knew about the conflict between Tumano and Ano Kare, but told that this conflict was solved. When these areas were visited, it appeared that there still were tensions between the two kebelles. Moreover, the woreda water bureau was not aware about the disputes within the kebelles Tumano. The officials could not tell the issues at stake and how the water committee there was dealing with it. Also the farmers building their own hand dug wells was something the woreda water bureau did not know was going on.

The water resources around which conflicts occur also differ between the woreda's. Some woreda's are actively developing groundwater systems, but in other woreda's this development is absent. It became clear that in all visited woreda's the irrigated area is increasing rapidly, with ambitious goals for the future. According to an irrigation expert of the Kobo woreda agricultural bureau, most conflicts in Kobo woreda happen around surface water systems. Around groundwater, there is less conflict at the moment. There are four-five disputes that reach the woreda level for assistance every year. The woreda officials will sit down with the involved parties and mediate to reach a solution. The conflicts have no influence on the functionality of the system (except in case a, Aybe kebelle). Surface water systems are more influenced by water shortages and are therefore more sensitive for conflicts than groundwater systems, although there is also a risk of over-exploitation here (Interview 29).

7.3 Lessons for local conflict management

Conflict resolution mechanisms differ per case and sometimes conflict resolution is almost completely absent. The goal of conflict resolution is to solve the issues at stake and transform the conflict situation in a situation of cooperation and sustainable management. It is important to realise that water management equals managing conflicting interests, especially around surface waters where domestic and productive used is often combined. Furthermore, conflicts can arise gradually from situations of void. Most cases are using discussion between the different stakeholders to solve conflicts, often by using woreda officials as mediator and party which can give advice (based on water availability and crop systems). Especially when two kebelles are sharing one water source or scheme, mediation of the woreda is needed for a distribution of the water. Most of the issues are solved on local scale. Also, study of the water resource is often part of the resolution process. In that way, experts can give a solid advice on how the water from the system should be used. The water committees are the first organisations to be responsible for conflict resolution. If the problems are beyond their capacity, the woreda level is asked for support. Local leaders and rules are sometimes used to mediate in conflict and dispute situations. The case Kombolcha (g) shows that bringing different parties together and using existing systems to do that helps in managing an irrigation system peacefully. In this case, a local NGO is intervening with success in creating an atmosphere were problems can openly be discussed. In most cases of void, there are some attempts in solving the issues at stake, but these attempts are not executed well or leading to concrete changes in management.

This chapter presented some examples of conflict management. In some of the other cases, there also have been attempts to solve or deal with conflicts, in others not. In the case of Aybe kebelle (case a), there have been attempts to solve the issues between the resistant group of farmers and the water committee. Several meetings have been organised to discuss the issues with the whole cooperative. However, the resistant group of farmers never showed up and encouraged other farmers not to come. The woreda agricultural bureau and the KGVDP also have tried to bring the different parties together. Now, the local government is withdrawing support, since they do not see how this issue can be solved. The chair of the water committee suggested that a change in the system could solve the issue. If the system would use furrow canals, the costs for operation and maintenance will be less than the case is now. However, changing the system is a costly action in itself (Interviews 3 & 4).

In the case of Raya-Azebo (case b), the local government involved in solving the conflict by studying the situation around the diversion and based on that, they made a distribution schedule. The government had a prominent role and the wishes and interests of the local communities were less important. The result is that this solution is not well received by the farmers; especially farmers from Genete kebelle are not satisfied with the solution of the woreda officials. They are dissatisfied for receiving less water than Tsigea kebelle. This reaction of the local government is not desirable. For more effective conflict resolution methods, an active and positive attitude of the local government is needed, most ideally as in the case of Kombolcha (case g), where the local government is acknowledging the potential for local conflicts and has an approach to deal with it (Interviews 12-14).

The distribution issues around the Kobo diversion are solved by using mediation of the local government and discussion between stakeholders of the two kebelles (Golina and Mengello). The water committees of the two kebelles are discussing the issues with each other under supervision and with advice of officials of the woreda agricultural bureau. This has led to a distribution schedule, but there is still some disagreement about it. Therefore the woreda officials are planning following-up meetings to ensure a good relationship between the two kebelles. This is a positive approach to the disagreement between the two kebelles. For the issue with the physical fighting incidence, the water committee used traditional systems, in this case the local court, to punish the offenders. The use of local courts is also mentioned during other interviews. These local courts exist of elderly men from the community who decide the punishment of the actions of the offenders. There are however no archives keeping records about the procedures and past decisions (Interviews 3 & 25).

One of the main points observed is the lack of attention for conflict resolution on the institutional level. Most woreda agricultural bureaus are in some way involved in local conflicts when the local committees ask for their help, but there is no conflict resolution procedure written down. Also in trainings to cooperatives members or woreda officials, there is no or little attention for conflict resolution. Particularly because the water committees are the first responsible organ for conflict resolution, it would be good to give attention to these kinds of things during training and establishment of the cooperatives. What can be observed from the cases is that solutions differ and how the solutions are defined. The role the local government is playing in these situations differs, as described in chapter 6. Not all solutions are positively received by the different stakeholders.

On the other hand, every situation is different. The contextual factors are also determining the characteristics of each situation. Therefore, training water committees and local government officials in blueprint solutions will not be the solution. It is important to create local awareness of the conflict, cooperation and void situation and make the water committees capable of handling these issues. It would also be good for woreda officials to have a step-wise plan to deal with conflicts. In these plans, it is good to involve local systems and practices in conflict resolution, because these will have the broadest support in the local communities. Furthermore, the sustainability of the actions is an important element to consider. As the Berki watershed case represents (case d) is that big project and attention does not always have to lead to long term changes and solutions for water management situations. After the project is over, people can go back to business as usual and the void situation can arise again, possibly leading to conflicts in the future.

Institutionalising of conflict resolution and prevention should be part of irrigation management. It would be good to train the water committees and make them aware about the possibility and diversity of conflicts. The local government should know how to deal with the issues and use a systematic approach, adapted to the local situation. Identification of the issues and interests is an important first point, followed by discussions between the different stakeholders. The local government can give advice about solutions (e.g. distribution schedules), but should be careful not to enforce solutions without acceptance of the local communities. The Ministry of Water and Energy can help by investing time in developing clear water law which defines water rights; clear water rights can contribute to the reduction of conflicts. They are however difficult to formulate and clarify. On lower levels, rules and regulations about water rights can also be specified; this is especially useful for river, since there are different types of uses there. Another important point for is the systematic record keeping of previous conflicts, to ensure that woreda officials can learn lessons from these cases and identify trends in their region. At this moment, due to lack of documentation and changes in positions, nobody is having an idea about the number and extent of local conflicts, especially on higher levels of government.

8. Conclusion

Investments in irrigation systems are crucial for improving the lives of the rural poor in Ethiopia and for transforming the agriculture sector to stimulate sustainable economic growth. Ethiopia has a large water resource potential to develop irrigation systems, but the development of this potential is going together with some major challenges, which are jeopardising the sustainability of the systems and can results in local problems and conflicts. This thesis presented some local management situations in Ethiopia with as aim to increase the knowledge of conflict and cooperation situations around irrigation systems. There is hardly documentation about these local water-related conflicts available. The central research question was: *What is the nature of situations of conflict, cooperation and void around irrigation schemes in Ethiopia, who intervened and what lessons can be learned from these situations?* To answer this question, seven cases of irrigation management in Ethiopia were presented in this thesis. These cases revealed that most conflicts and disputes are very local, between farmers using the same scheme, between the farmers and the water committee or between farmers from two different kebelles using the same water resource. Only the Berki watershed (case d) represented a conflict on a broader scale, involving a river shared by two woreda's.

The cases have been presented in a continuum from conflict to cooperation. However, that raises a question: when is a conflict a conflict? In all management situations, there were disputes and problems present or have been present in the past, but not all situations represent a conflict situation. In the theoretical chapter, a conflict situation was defined as follows: *"open competition for the water resource, where conflicts and tensions are arising between users, managing institutions or the (local) government. The competition is leading to a negative influence on the water system, but does not necessarily have to be violent"*. The last factor is important, but makes the difference between a void or conflict situations sometimes difficult to assess. Therefore, the fact of negative influence on the water system is important. Most of the visited cases represented situations were problems and disputes were present, but these problems were not escalating into (violent) conflict at the moment of the fieldwork. Farmers are often not satisfied with the water distribution or the influence of the local government on their farming practices. And it is in a lot of cases difficult for the water committees to control the water flow, due to the use of inefficient canals to lead the water to the different fields. However, if issues/disputes between farmers are not taken seriously and are not settled on time, more (violent) conflicts could arise, especially with the rapid developments in irrigation.

The occurrence of conflicts appeared to be closely related with the time dimension, connected to the climatic conditions and the water availability throughout the year. Since the farmers still use rain as irrigation during the main rainy season (June-September) and the water sources are depending on the same rainfall for their recharge, the occurrence or absence of rain can have an influence on the number and intensity of conflict situations. In general, there are more conflicts and problems signalised and reported to the woreda offices at the end of the dry season. Especially when the small rainy season around March is absent, the number of local conflicts will increase. This time dimension is enforced by the finding that these management situations are dynamic and can change of nature over time. Void situations can become conflict situations at the end of the dry season, when there is not enough water left to supply water to all the farmers. Attempts to solve problems and conflicts are also contributing to the change of nature; a conflict situation can be transformed into a situation of cooperation.

This fact is bringing the discussion back to scarcity, or with other words, the availability of water throughout the year. However, this scarcity can be classified as economic scarcity and could be overcome by good functioning institutions that can adapt the water demand in times of scarcity (e.g.

by using different crops). This point is related to the claim of Homer-Dixon, that scarcity alone is not causing conflicts, it is in interaction with other factors that conflicts arise around water systems (1999). Therefore it can be said, that water availability due to rainfall is influencing the occurrence of local conflicts, but in combination with other factors, like distribution and functioning of institutions. This scarcity can trigger problems that already exist around irrigation systems and transform cooperation and void situations into conflicts. And this scarcity can be overcome by technical innovation (making irrigation systems more efficient) and cooperation (effective institutions), like Boserup claimed (1965). The institutional site is important to develop in the future, to make irrigation systems a positive development in Ethiopia.

This time dimension as such is not mentioned in the literature, therefore it has not been identified in the theoretical chapter. It is however an important element to take into account during irrigation management. If the local government and water committees are aware about the higher risks of conflicts during the dry season or when the rains are delayed or less active, these conflicts can be identified in an early stage, or precautions can be taken to lower the water demand temporarily (for example by choosing different crops or limited the irrigated land). Maybe the time dimension is leading to temporal scarcity. This factor is also contributing to the dynamics of the local situations; the situations change over time of their nature. A situation of conflict can return to void or cooperation, but after a few months/years new problems can arise or old tensions can cause trouble again.

When looking back at other the factors causing different situation, the research revealed that functioning of institutions is an important factor, which is also influencing other factors like distribution and competition of water. This is in line with the theoretical findings. Local water committees can for example make a clear distribution schedule, which can help in minimising the risk for problems. And most water committees need support from the woreda organisations in how to manage the systems and how farmers are supposed to use the land. Around irrigation management, various government institutions are involved which need to work together to support the water committees and construct new irrigation systems. These organisations need to work together well, which is not always the case now.

Local conflict resolution is often based on discussions with stakeholders involved in the situation, without guidelines to lead this process. The woreda government is often asked to mediate in the problem by the water committees. The role the government is playing in the local management situations differs from case to case. Some woreda officials play a supporting role, by bringing different parties together and constructively work on a solution, while other woreda officials are enforcing their solution on the situation, often resulting in resistance from some of the farmers. On higher level, the government is pushing the development of irrigation systems in Ethiopia rapidly. On every level you see big differences in irrigated areas last year, this years and ambitious plans for the next years. It is clear that irrigation development is one of the key priorities of the Ethiopian government, not only for large scale farmers, there is also a focus on small-scale farmers and systems. In this development, the focus seems to lay on technical issues and the construction of new systems. After the systems are operational, less attention is paid to the management and social aspects. As a result, different kind of problems may arise, which can jeopardise the sustainability and functionality of the systems. In the case of deep boreholes, the borehole itself is constructed for a lot of money, but an efficient system to distribute the water from the borehole to the fields is sometimes missing. Also, new cooperatives experience some starting issues, which could lead to conflicts when these are not addressed. And from different researches (also from this one) it is concluded that water governance and management is essential for the long term functionality of the systems.

During the fieldwork, various types of irrigation systems have been visited and some differences have been observed in the management of the systems. The conflicts around groundwater systems are generally about the distribution and not about the availability of water. Available water is located underneath the ground and is not that visible for the local farmers. In the case of surface water, it can also be about the distribution, but mostly about water availability and scarcity, especially the division of water between upstream and downstream users. Surface water resources are more often shared between two kebelles or woreda's, which is often a cause for discussion about the division of water. During the fieldwork, there were more conflicts observed around surface water systems and this was also confirmed by some woreda officials. This is caused by the fact that surface water resources are used more often for irrigation system and the groundwater developments have been recent (in some woreda's almost absent). However, groundwater conflicts are expected to arise more when more systems have been developed.

Ethiopia can be seen as a developmental state, which is influencing the pathway to development for the country. The Ethiopian government has a strong say in the developments in the country. There is not a lot of room for opposition, which often means that local parties can influence the politics in the woreda's and kebelles. It is very likely that important and respected people from the kebelle are chosen in local water committees, enforcing local power play in the water committees. Therefore, the political and social relations are important in the local management situations. In recent years, Ethiopia is developing rapidly as a country, making big jumps in economy and other sectors. When looking back at the role of the government in irrigation systems, the government is responsible for the construction of irrigation systems. This approach is helping Ethiopia to develop irrigation system on a rapid scale, but seems not really though out. There is no room for individual development and there it too little attention for the management part which is coming into place after the construction. The developmental approach of the government is leading to more irrigation development on the one hand, however, this is also leading to some local problems where the federal government has no knowledge about.

8.1 Discussion

This master research has revealed several interesting insights in the way small scale irrigation systems are managed in Ethiopia. When reflecting back on the conceptual framework (figure 3.1), the theory was for the most part in line with the empirical findings. The prominent place of institutions in the occurrence of conflicts and in conflict prevention and resolution was visible in the different empirical cases. However, the importance of the time dimension is missing. The local management situations are dynamic and can change over time. Time dimension should be a factor in the conceptual framework as well. This factor is closely related to the factor economic scarcity, since in dry periods, water may be scarce in some systems, but this can be overcome by good management and the development of more systems. The factor physical scarcity was not really an issue in the visited cases, since in all visited areas, there was a potential to develop more (ground)water systems for irrigation. In some other parts of Ethiopia, this physical scarcity might be an issue. The research has shown the importance of management and the need to invest time and money in this issue. Just building new system without much attention for management and conflict prevention/resolution is not leading to a sustainable result.

When looking back at the research process, it can be noticed that a lot of data was gathered in a relative short period of time. This research chose to visit different cases in three regional states. This meant a lot of travelling and limited time per case. This approach was chosen to get an impression of the variety of management situations. However, repeated visits over a longer period of time would have been interesting, especially since the research shows the importance of the time dimensions in the number of situations becoming conflict situations and the dynamic nature of the management situations. The repeated visits would also give to opportunity to win the trusts of the different interviewees and have time to visit all relevant stakeholders in a case. For a master research, this is however difficult to realise.

Timing issue also influence on situations I visited. The beginning of my fieldwork was just before the small rainy season, but at the end of my first fieldwork period, the first rain drops started to fall from the sky. During the second fieldwork period, the small rainy season was started in various parts of the country, however not in the area of my second fieldwork period (Degem). During my third fieldwork period, the rainy season has caused a problem in the irrigation system I was visiting, but the same rain made it no problem for the farmers; they could continue with their production. It seemed like a good year for the farmers, since it was not really dry between the small and second rainy season, at least not in Addis Ababa, where it started raining in the half of March. Various people in Ethiopia told me that last year, this small rainy season was absent; a potential for more local conflicts, although no concrete data has been found about the occurrence of conflicts last year.

During the research process, a lot of data has been gathered, mostly due to semi-structured interviews with different stakeholders in the local situations as well as regional and national government organisations. The availability of written sources was more difficult. It was tried to collect these sources, but they were not always reliable or comparable between cases. Therefore, these sources have been used carefully. It appeared that the contextual factors played an important role in the occurrence of problems around irrigation systems and how these problems are developing over time. Ethiopia is a country with major regional differences in climate conditions as well as in development. The approaches of different regional governments are different. However, these factors have not been closely investigated during the research period.

8.2 Recommendations:

From this research, different lessons can be learned. Here, lessons for irrigation developments in Ethiopia and more general recommendations for further research are presented.

8.2.1 Recommendations for stakeholders in Ethiopia

To improve the situation around small scale irrigation system, some lessons can be learned from this research. Most of the lessons are for the (local) government, but could also be used for NGOs working in the irrigation sector in their approach and training programmes.

- It is important that the woreda officials (or even regional officials) know that disputes are likely to happen in situation where a water sources is shared by different kebelles (or woreda's). Attention for this issue during the construction of system and early mediation of the woreda officials is could avoid these kinds of tensions and conflicts.
- Related to the first recommendation is the establishment of a system based water committee in cases of two kebelles sharing a water source of system can help to manage the shared system better. Now a lot of water committees are kebelle based, representing the farmers using the system in that kebelle. When there is one committee per system or water resource, there are not two water committee fighting with each other over the water distribution, but is one committee responsible for the distribution. This might work to manage the shared systems better. The plan is to introduce one such a water committee in the diversion in Kobo (case f). If this works, this could

be a good solution for other shared diversions as well. Case c (two kebelles sharing a deep borehole) represent that shared committees not always work, but if well overthought, it might be a solution.

- Develop conflict management guidelines for the different regional states (the regional states have slightly different organisation structures). Train local government officials and water committee members in conflict management tools, how to signalise potential conflict and deal with different kind of situations. Map the different risk per irrigation system to be aware about the possible scenarios.
- Every irrigation scheme is different. Local government officials should recognize the complexity and diversity of managing water resources in agriculture. There is no one-size-fits-all solution for preventing and solving conflict situations and prevent any problems from arising.
- Pay attention to access rights to water resources and property rights of small-scale farmers. Especially for young farmers it is difficult to get access to land and the related water resources. They are often dependent on renting land for older people with land access. But this creates an insecure situation for them, because they can only rent the land for one season at a time and during the rainy season, the land owners often want to use the land for themselves.
- Improve policies and regulations for irrigation systems, with focus on water fees, water rights, water conflict resolution and cooperation between different governmental levels. Water rights are for example not regulated; everybody can take water from the rivers. In Tanzania for example, there are water basin boards that regulate access to rivers. This can be a possibility for Ethiopia as well (Interview 9).
- Pricing of water is a vital issue. Especially for surface water schemes it is important to enforce water pricing. In this way, people will be more aware about the value of water and will use the water more efficient. The money can be used for maintenance issues and the water committee will feel more ownership over the systems. Most of the systems have user payment, but in the ones that did not had payment, this appeared to be a problem.

Improve the cooperation between the different (local) government institutions working on water and irrigation. There are now different governmental organisations involved in irrigation on the woreda level. It is good to look at the organisations involved in irrigation developments and invest in improved ways of cooperation or a better grounding of irrigation in exiting organisations (for example by creating an irrigation department within a woreda agricultural bureau). The Oromia regional state is planning to change the regional organisation, to establish a regional irrigation bureau. Whether they will succeed in this plan, have to be seen. Again, irrigation remains a tricky subject, since water and agriculture are linked. Moreover, in cases of multiple-use, especially rivers and other surface source, the water bureau has to be involved.

- Keep investigating the potential sources for irrigation. In a lot of areas, (ground)water is available, but not widely used; there is a large unused potential in Ethiopia. And irrigation systems can mean a real improvement of the local population. The government should keep oversight over the situation and invest also in the environmental aspects, to make sure water will be available in the future as well. Knowledge training for local government officials, they often lack practical experience. The people working in the kebelles and the woreda's are often fresh graduates who have a lot of theoretical knowledge, but lack the practical skills.

8.2.2. Recommendations for further research

The research contributed to more knowledge about local management situations in Ethiopia, but there are still some elements that need further research for better understanding. Therefore, the following recommendations for further research have been formulated.

- It would be good to conduct a research with more long term observations around a small number of cases. With repeated visits to one case, the dynamics around management situations can better be understood, as well as the development over time, which was identified as important element in local conflicts. As has been seen in this research, local management situations are dynamic and can change over time due to different factors. It is desirable to understand these dynamics better. With repeated visits, there is also more room to talk the all different stakeholders involved extensively. And it is important to create room to talk to the different stakeholders without the presence of other stakeholders, e.g. talk to the farmers without presence of woreda officials.

- This research only focused on small-scale conflicts where smallholders are involved. In some woreda's there are also interesting issues with private farmers and small-scale farmers (Raya-Azebo, Alamata and Kobo). Sometimes, problems between the private farmers and small-scale farmers can turn into conflicts. I even heard rumours about farmers burning down part of the land of a large private farmer in Alamata woreda. These issues are probably related to land ownership and the relation between the private farmers and the community, but it would be interesting to further look at this issue (Interview 41). The University of Mekelle will also be involved in private landownership in Alamata woreda for educational purposes, but it is an interesting development to investigate; What is the influence of these kind of developments on the local farmers, and what happens when farmers have to leave their land for the establishment of a private farm, what are their rights in these resettlement processes?

- Local organisations institutions are an interesting element to further investigate. Besides local organisations, traditional unwritten rules, behavioural norms and local power relations are also influencing the background against which local water-related conflicts are arising. Therefore, it is important to pay attention to the broader definition of institutions, especially on lower level, since these are closed related to the conflicts and are less documented. It would for example be interesting to making an in-depth case study of good functioning local organisation managing water resources, for example the Gadaa system of the Borena people in the Southern part of Ethiopia (textbox 4). This system is complex, but represents a system which has been working for a long time to manage the natural resources of the Borena people. This would give a good example of the influence of local organisations in a country with strong local tribes and communities. A note has to be made here, that this example is not representative for all parts of Ethiopia, but it will be an interesting example of the strength of local organisations.

- Investigate the environmental impact of irrigation systems. With means of irrigation, the land is used more intensively, resulting in more use of fertilizer and pesticides. These elements can degrade the land, but also reach the groundwater table and pollute this resource. It would be good to analyse the influence of the new agricultural practices on the land and water resources and give recommendation to make the environmental impact as low as possible.

Closing statement

This thesis has shown the importance of attention for local water-related management situations. Ethiopia is investing in irrigation developments for small scale farmers, without paying much attention for what happens after the construction of the new systems. There is still a lot of unused (ground)water potential which can improve the livelihoods of a lot of farmers who still struggle to survive. Improvements in the agricultural sector mean improvements in the developing country as a whole. Ethiopia is a country where a lot of people are depending on food security. Irrigation can reduce this dependence a lot and increase the health and education of households. Water is life and agriculture the main source of income. Therefore irrigation can be a strategic element in the development strategies which countries can use. It is important that the water resources are used on a sustainable way, meaning that the resources are not degraded or over-exploited and problems are solved and prevented. Investments in irrigation systems can mean a sustainable push in development, when there is enough attention for all sides of sustainable irrigation development; this means attention for the management and other social issues, besides the attention for the environmental side.

Literature list

ADSWE (2011) Brochure about the Amhara Design and Supervision Works Enterprise. ADSWE, 2004 Ethiopian Calendar, converted to 2011 Gregorian calendar.

Annan, Kofi (2002, February 26). World's water problems can be 'catalyst for cooperation' says Secretary-General in message on World Water Day [Press release]. http://www.un.org/News/Press/docs/200 [Last reviewed on 16-07-2013]

Arsano, Y., Mekonnen, E., Gudisa, D., Achiso, D., O'Meally, S., Calow, R. and Ludi, E. (2010) Governance and Drivers of Change in Ethiopia's Water Supply Sector. A study conducted by the Organisation for Social Science Research in Eastern and Southern Africa in collaboration with the Overseas Development Institute (ODI), May 2010.

Awulachew, S.B. (2010) Irrigation potential in Ethiopia: constraints and opportunities for enhancing the system, IWMI, July 2010, with contributions of Erkossa, T. and Namara, R.E.

Boserup, E. (1965) *The conditions of agricultural growth: the economics of agrarian change under population pressure.* London, George Allen and Unwin LTD, Ruskin house museum street.

Butterworth, J., Visscher, J.T., van Steenbergen, F. and van Koppen, B. (2011) Multiple use water services in Ethiopia: scoping study. A report for International Water Management Institute (IWMI), International Water and Sanitation Centre (IRC) and the Rockefeller Foundation, December 2011.

Carius, A., Dabelko, G.D. and Wolf, A.T. (2004) Water, Conflict, and Cooperation. Policy brief, the United Nations and Environmental security, ECSP report, issue 10.

CIA (2013) World Fact Book Ethiopia. https://www.cia.gov/library/publications/the-world-factbook/geos/et.html [Last reviewed on 28 May 2013]

Cold-Ravnilde, S.M. (2012) A gift from God: conflicts over water and authority in Mali. PhD Thesis – University of Roskilde 2012.

Costantinos, B.T. (1998) Post Cold-War environmental governance and Alternative Natural Resource based Conflict Management (ANRCM): processual and strategic dimensions to conflict management in the African political transition period and the Boran case study. Background and resource paper for the regional conference on ANRCM, June 2-3 1998, United Nations conference centre, Addis Ababa, Ethiopia.

ECWP (2007) Ethiopia Country Water Partnership (ECWP): promoting and implementing Integrated Water Resources Management (IWRM) in Ethiopia. Background document prepared for Tigray Regional State key sector officials consultative meeting, May 26 2007 Mekelle.

Edossa, D.E., Awulachew, S.B., Namara, R.E., Babel, M.S. and Gupta, A.D. (2007) Indigebous systems of conflicts resolution in Oromia, Ethiopia. Chapter 9 in: Van Koppen, B., Giordano, M. and Butterworth, J. (2007) *community-based water law and water resource management reform in developing countries.* CABI International, 2007.

Endalamaw, A.M. (2009) Optimum utilization of groundwater in Kobo Valley, Eastern Amhara, Ethiopia. A thesis presented to the faculty of the graduate school of Cornell University.

FDRE (2013) Government overview. Federal Republic Ethiopia, Ethiopian government portal. http://www.ethiopia.gov.et/web/Pages/GovernmentOverview [Last reviewed on 20-05-2013]

FDRE (2010) Definition of Power and Duties of the Executive Organs of the Federal Republic of Ethiopia. Federal Negarit Gazeta of the Federal Republic of Ethiopia, proclamation number 691/2010, year 17, number 1, Addis Ababa October 2010.

FDRE (ND) Berki Watershed Development Plan. Federal Republic of Ethiopia, developed for the pilot project of the ECWP, no date.

Funder, M., Bustamante, R., Cossio, V., Huong, P.T.M., Van Koppen, B., Mweeda, C., Nyambe, I., Phuong, L.T.T. and Skielboe, T. (2012) Strategies of the Poorest in Local Water Conflict and Cooperation – Evidence from Vietnam, Bolivia and Zambia. *Water Alternatives,* 5(1), pp.20-36.

Funder, M., Mweeba, C., Nyambe, I., Van Koppen, B. and Ravnborg, H.M. (2010) Understanding local water conflict and cooperation: The case of Namwala District, Zambia. *Physics and Chemistry of the Earth,*

35(2010), pp.758-764.

Goodhand, J., Vaux, T., and Walker, R. (2002) Conducting Conflict Assessments: Guidance Notes. Department for International Development (DFID), London.

GoogleMaps (2013) Maps of Ethiopia, places I visited. https://maps.google.nl/maps?hl=en&q=afleiding [Last reviewed on 18-05-2013].

GW-MATE (2011) Ethiopia: Strategic Framework for managed groundwater development. GW-MATE (Groundwater Management Advisory Team) in cooperation with Ethiopian Ministry of Water Resources, MetaMeta, Acacia Water and Nuffic.

GWP (2010) Transmitting tensions down the river: How to resolve them. Policy brief – a case study of Ethiopia IWRM Implementation Pilot Project, Global Water Partnership (GWP) and Ethiopia Country Water Partnership, 29-06-2010.

Harvey, P. A., & Reed, R. A. (2006) Community-managed water supplies in Africa: Sustainable or dispensable? Community Development Journal, 42(3), pp.365-378.

Henze, P.B. (2000) Layers of Time. A History of Ethiopia. Hurst & Company, London.

Hodgson, G.M. (2006) What are institutions? Journal of economic issues, vol XL (1), pp.1-25.

Homer-Dixon T (1999) Environment, Scarcity, and Violence (Princeton: Princeton University Press).

Kebede, S. (2013) Groundwater in Ethiopia: features, numbers and opportunities. Springer Hydrogeology, Heidelberg 2013.

Malley, Z. J. U., Taeb, M., Matsumoto, T., & Takeya, H. (2009) Environmental sustainability and water availability: Analyses of the scarcity and improvement opportunities in the Usangu plain, Tanzania. Physics and Chemistry of the Earth.A/B/C, 34(1-2), pp.3-13.

MoFED (2010a) Ethiopia: 2010 MDGs report. Trends and prospects for meeting MDGs by 2015. Ministry of Finance and Economic Development, September 2010.

MoFED (2010b) Growth and Transformation Plan 2010/11-2014/15. Volume I: main text. Ministry of Finance and Economic Development, November 2010, Addis Ababa.

MoWR (1999) Ethiopian Water Resources Management Policy. Ethiopian Ministry of Water Resources, 1999.

MoWR (2001) Ethiopian Water Sector Strategy. Ethiopian Ministry of Water Resources, 2001.

MoWR (2002) Water sector development program. Main report volume II. Ethiopian Ministry of Water Resources, October 2002.

OECD (2005) Water and violent conflict. Issue brief Development Assistance Committee (DAC), mainstreaming conflict prevention, OECD, 2005.

Ohlson, L. (2000) Water conflicts and social resource scarcity. Physics and Chemistry of the Earth,25(3), pp.213-220.

Ostrom, E. (1990) Governing the Commons: The Evolution of Institutions for Collective Action. Cambridge: Cambridge University Press.

Penning de Vries, F.W.T. (2007) Learning alliances for the broad implementation of an integrated approach to multiple sources, multiple uses and multiple users of water. Water resource management, 21(2007), pp.75-95.

Perry, J.A. (1996) Water Quality: Management of a Natural Resource. John Wiley & Sons, United Kingdom.

Ravnborg, H.M., Bustamante, R., Cissé, A., Cold-Ravnkilde, S.M., Cossio, V., Djiré, M., Funder, M., Gómez, L.I., Le, P., Mweemba, C., Nyambe, I.; Paz, T., Pham, H., Rivas, R., Skielboe, T. and Yen, N.T.B. (2012) The challenges of local water governance: The extent, nature and intensity of local water-related conflict and cooperation. Water Policy, 14(2012), pp.336-357.

Ravnborg, H.M. ed. (2004) Water and conflict: conflict prevention and mitigation in water resource management. Danish Institute for International Studies, DIIS, 2004:2

Savenije, H.H.G. and Van der Zaag, P. (2008) Integrated water resource management: Concepts and issues. Physics and Chemistry of the Earth, 33(2008), pp.290-297.

Schlein, L. (2013) Ethiopia: Hunger Costs Ethiopian Economy Billions of Dollars. Voice of America (Washington DC), 25 June 2013.

Sokile, C.S. and Van Koppen, B. (2004) Local water rights and local water user entities: the unsung heroines of water resource management in Tanzania. *Physics and Chemistry of the Earth,* 29(2004), pp.1349-1356.

Sørenson, P. N. and Bekele, S. (2009) Nice children don't eat a lot of food: Strained livelihoods and the role of aid in North Wollo, Ethiopia. Forum for Social Studies, Addis Ababa, Ethiopia.

Sreeramulu, U.S. (1998) *Management of water resources in agriculture.* New Age International publishers, New Dehli. Reprinted version from 2005.

Tafesse, M. (2003) Small-scale irrigation for food-security in sub-Saharan Africa. Report and recommendations of a CTA study visit, Ethiopia January 2003. CTA working document number 8031.

Tucker, J. and Yirgu, L. (2010) Small-scale irrigation in the Ethiopian highlands: what potential for poverty reduction and climate adaption? Ripple briefing paper, number 3, August 2010.

Turner, M.D. (1999) Conflict, Environmental Change, and Social Institutions in Dryland Africa: Limitations of the Community Resource Management Approach. *Society & Natural Resources,* 12(1999), pp.643-657.

UNDP (2013) General maps, administrative maps of Tigray, Amahara and Northern Oromia regional state. UNDP Eergencies Unit, Addis Ababa. http://www.africa.upenn.edu/eue_web/menu4596.htm [Last reviewed on 6-06-2013].

UNDP (2012) Ethiopia Country Profile: Human Development Indicators http://hdrstats.undp.org/en/countries/profiles/ETH.html [Last reviewed on 17-12-2012].

Van Koppen, B.; Smits, S.; Moriarty, P.; Penning de Vries, F.; Mikhail, M. and Boelee, E. (2009) Climbing the water ladder: Multiple-use water services for poverty reduction. Technical Paper Series No. 52. The Hague, the Netherlands: IRC International Water and Sanitation Centre and International Water Management Institute.

Van Steenbergen, F. (2011) The Politics of Conflict, Cooperation and Void in Groundwater. Main theoretical framework for the CoCoon project.

Van Steenbergen, F. and Verheijen, O. (2006) Working on spate irrigation: lessons from the field. Chapter 19 in: Van Lindert, P., De Jong, A., Nijenhuis G and Van Westen, G (red) , Development matters, geographical studies in development processes and policies. Faculty of Geosciences, Univerisy Utreht, pp.283-291.

Verheijen, O. (prepared document) (2011) Sustainable Water Harvesting and Institutional Strengthening in Amhara (SWHISA) Project: procedure manual for establishment and capacity building of water users' organisations in Kogo irrigation and drainage system. Hydroconsult Inc Canada, in association with Agrodex-Oxfam Canada, Bahir Dar, October 2011.

WaterAid (2009) Management for Sustainability: Practical lessons from three studies on the management of rural water supply schemes. WaterAid Tanzania, June 2009.

Wengert, N. (1983) Water Management Institutions. *Operation of Complex Water Systems,* NATO ASI Series, Volume 58, 1983, pp.277-292.

WFP (2013) Food security Ethiopia. http://www.wfp.org/countries/ethiopia/food-security/overview [Last reviewed on 6-05-2013].

XE currency convertor (2013) Exchange rate Ethiopian Birrs to Euros. [Last reviewed on 29-04-2013] http://www.xe.com/currencyconverter/convert/?Amount=1&From=EUR&To=ETB.

Appendix I Question lists

This appendix present the basic question lists used during the research for interviews with woreda officials from the water and agricultural offices, farmers, members of local water committees, officials from regional water and agricultural offices and interviews with NGOs/knowledge institutes. The question lists form the basis and were adapted to the specific interview situations.

Question list woreda water/agricultural bureau
General information: date of the interviews, people present (+ function),
Woreda name, number of kebelles

Rural water supply
1. How is the rural water supply organized in this region?
   - Responsibilities of the water bureau regarding irrigation and drinking water
   - Responsibilities of the agricultural bureau regarding irrigation
   - Cooperation between the water and agricultural bureau
   - Cooperation between the woreda, zone and region
   - Organisation of this bureau (processes/department, number of staff)
2. What kind of systems are used for drinking water and for irrigation?
   - Combined or single use systems?
3. How are the systems managed? On what level (community)?
   - Are the committees collecting fee and making rules of usage?
   - How are committees established?
   - Responsibilities of the committees?
   - Number of cooperatives/ water committees?
4. What has been achieved with regards to irrigation and drinking water?
   - Major changes in the last 5 years?
   - Irrigated area: this year and last year
   - Drinking water coverage (urban/rural)
   - Major differences between kebelles?
5. What are the regional government's water service priorities?
   - What are policy targets?
   - What is the plan for coming year
   - How are the Woreda Development plan and/or Woreda WASH plan developed?
   - Documentation?
6. How is the budget made for the woreda agricultural/water bureau?
   - What criteria are used in allocating budget for the water sector?
   - Are kebelles/communities involved in the budget process, e.g. by making request to the offices for new schemes?
   - What are the sources of the money? (Federal, NGO etc.).
   - What support does the region/district receive from the National Government, NGOs and other donors? (financially and technically)
   - Responsibilities of different levels/organisations (state, region, woreda, kebelle, other organisations?
7. How is the woreda water/agricultural bureau overseeing the performance of the water schemes in the woreda?
   - What monitoring mechanisms and reporting systems are available? Field visits, development agents?
8. How is land ownership arranged in this region/district etc.? And the ownership/rights to use water?
9. Are there (foreign) private investors present in this woreda?

- If yes, how is the relation between the agricultural/water bureau and these investors?
- Do they get support from this office? Control mechanism?
- Resettlement programmes?

10. How is relation between the woreda water/agricultural bureau and the local population?
- What mechanisms are in pace for citizens' involvement? Needs assessments, participatory planning processes?
- How is the relation between the agricultural bureau and the farmers? On what basis get the farmers information about agricultural inputs? How organised?

Conflicts/problem/challenges

11. Why kind of problems are happening around water systems?
- What is the nature of these problems? Social, technical, socio-sanitation?
- Causes of the problems? Related to water availability?

12. Do these problems have a negative influence on the functionality of the water systems?
- Do these problems negatively influence the local population? What is the impact on the local population?

13. Who are involved in the problems?

14. What is role of the woreda water/agricultural bureau in the problems?
- Kebelles, local institutions, local leaders?

15. How it developed
- Who/what usage prevailed
- Role/no role of traditional mechanisms, local administration/ local courts/ local party politicians or higher up organisations
- Actions taken
- Types of conflict resolution mechanisms, locally and on woreda level?
- Impact on the local population

16. Is there data about problems? Are the problems reported? Data available? Local courts involved?

17. Main challenges for the future? Possible scenarios for conflicts over water?

Question list farmers

General:

1. Name, age, household size
2. Profession

Water usage:

3. Land size of the farmer, types of crops to grow, livestock?
4. Water used for:
- Domestic usage
- Irrigation
- Livestock farming

5. How often en how much water is used for the different purposes?
6. Are you using 1 water source, or more? What kind of source/ irrigation system?
- Privately owned or public (sharing with how many)?
- If privately owned: how financed? How maintained? Why decided to invest in an irrigation systems?
- Satisfied about the water systems in use?

7. How would you describe the water situation in this community?
8. How would you describe the availability of water in this community?
- Are the water sources in the village sufficient for your daily activities?

- Are the water sources in the village sufficient for irrigating your land?
9. How are these water source managed?
10. What are the rules around water points?
11. Are you satisfied with the water resource system management in this community?
12. Crop selection/rotation? Link with the agricultural bureau? Fertilizer (biological)
13. Livelihood improvements since irrigation scheme is operational? Food security/ money

Water problems:
14. What are the major problems or inconveniences the community is facing with the water systems? If yes, what kind of problems, describe the problems. Causes, impact, effect on the community, water system.
15. What are your main concerns regarding the water situation in this community?
16. Do you contribute to the maintenance of the water systems?

Link with local government:
17. Do the people in the community feel like they are the priority of the (local) government concerning the availability of water?
18. Are there any ways the community can communicate their concerns about water availability and problems with the water systems to the government?
19. Are there recent examples of communications with the local government? If yes, did it help?
20. Is there any kind of organisation within the community with the intention to solve problems with the water systems?
21. Have there been any improvements concerning the water conflicts in the past few years?

Question list farmers' cooperative – water committee
Present + data + name cooperative + function in cooperative
Kebelle?

1. What kind of water/irrigation scheme is this cooperative managing?
   - Characteristics? Depth of system, way of distributing water?
   - Litre per second? Operation hours?
   - Year of construction?
   - For what uses is the water used?
   - Hectares irrigated, drinking water etc.
   - Farmers depending on it?
2. How is the system managed? Daily business?
   - How is the water distributed over the land? How schedules are made, who is responsible?
3. How many members has the cooperative? And the committee responsible for the management of the scheme? Gender balance? Volunteers?
4. Are the members of the cooperative elected or appointed? How does the election work?
5. What are the responsibilities of the committee?
   - What are the roles of the different members?
   - Also involved in agricultural inputs?
   - Did the members get training? From who?
   - Is the committee training the farmers on how to use the irrigation system/ agricultural inputs?
6. Is there a constitution or rules of usage? How developed?
   - Fines for stealing/cheating/disobeying?
   - Is the cooperative registered at the Woreda Agricultural Bureau?
7. Are there problems with functionality of the scheme? Availability of water? (causes)

8. How is the maintenance of the scheme arranged? Protection of the scheme?
9. Are the problems around the cooperative itself?
10. Are there other institutions involved, for example the woreda government? Support from the local government?
11. Who paid for the construction of the water point?
12. Do members contribute money to the cooperative? What amount? Special arrangements?
13. Does the cooperative have a bank account for the money?
14. Is the cooperative aware about problems around water systems in this woreda? Is there competition for water resources?
15. Main challenges/problems?
    - How are these problems dealt with/solved?
    - Role of committee/government/traditional leaders?

Question list regional water/agricultural bureaus

General information: date of the interviews, people present (+ function), region

1. How is the rural water supply/irrigation organized in this region?
    - Performance of the water systems: what is the current status of service delivery (water coverage or access figures drinking water and irrigation) by woreda (urban/rural)?
    - How is the regional water board supporting the functionality of the systems? Major regional projects?
2. What kind of systems are used for drinking water and for irrigation?
    - Cooperation between regional water and agricultural bureau on irrigation?
    - Support to woreda bureaus – monitoring mechanisms
3. What are the regional government's water service priorities?
    - What are policy targets?
    - Documentation?
4. How much of the regional budget is allocated to water supply (both drinking water and irrigation)?
    - What criteria are used in allocating budget for the water sector?
    - What are the sources of the money? (Federal, NGO etc.). Technical support?
    - Are local stakeholders involved in the process of planning of water projects/programmes? Who is involved?
5. What major changes have taken place in the past 5 years? Is the situation improved?
    - What are socio-economic, political, cultural and historical characteristics that influence the water sector performance?
6. What are the plans for the coming years?
7. Are there foreign large-scale farmers present in this region?
    - Policy towards private investors?
8. How is relation between the regional authorities and the local population?
    - How does the region respond to the needs of various segments of the population? (E.g. poor, women, children, disabled, pastoralists). Who is involved?

Conflicts/problem
9. Why kind of problems are happening around water systems?
    - What is the nature of these problems?
10. Do these problems have a negative influence on the functionality of the water systems?
    - Do these problems negatively influence the local population? What is the impact on the local population?
11. Who are involved in the problems?

12. What is role of the Regional water department in the problems?
    - What is the role of Woreda water offices in the problems?
    - Kebelles, local institutions?
    - Maintenance issues
13. How it developed
    - Role/no role of traditional mechanisms, local administration/ local courts/ local party politicians or higher up organisations
    - Actions taken / Impact on the local population
14. Is there data about problems? Are the problems reported? Data available?
15. Ambitious and rapid development in irrigated area. How is the regional government overseeing this development and making sure the development is having a positive influence on the local population?
16. Role of regional government in deep borehole design, construction and maintenance? How organised? Selection criteria? Problems in maintenance?
17. Giving training to the farmers or woreda bureaus? What kind of training?
18. Role of the regional government in watershed management?

Question list NGO/knowledge institute (REST, ISD, Water Action, World Vision and ADSWE)

1. What are the main tasks of the organisation?
    - Main fields or work
    - Main regions of projects/research
2. What is the organisation doing in the field of water (management) → drinking water and agriculture systems?
    - Drilling wells? How to determine where to drill wells?
    - Integrated Watershed Management Programmes
    - Agriculture → irrigation
    - How does the organisation help the small scale farmers in practice?
    - What about ownership of the projects/schemes? Are the local people contributing to the projects/schemes?
3. What are the responsibilities of the organisation, especially regarding water issues?
    - Have these responsibilities changed in recent years?
4. How does the organisation determine where to implement their projects/ water schemes? Are there special criteria?
5. How does the organisation cooperate with the government (different levels)
    - And with the other offices of the organisation?
    - And with local communities/organisations?
    - How is the contact with the local communities?
    - Following up activities after projects?
6. How is the organisation funded? What are the main sources?
7. What are the major developments in drinking water and irrigation? What is the role of the organisation in these developments?
    - What is in the planning for the coming years?
8. Is the organisation aware of local conflicts around groundwater systems (especially irrigation systems)?
    - If yes, how does the organisation deal with these local conflicts?
    - What is the nature of these local conflicts? Technical, social, economic?
    - Are there conflict management and resolution mechanisms in place?
9. Is there data available on local water supply/organisation and local water conflicts/competition?
    - Other relevant data?

Appendix II Interviews per case

The table below present the interviews held per case. Besides this case specific interviews, the interviews with Regional Officials and with the Ministry and other institutions in Addis Ababa provided a background for the whole research.

Table II: interviews per case study

Case	Type of interviews	total
A: Aybe kebelle, Kobo	- KGVDP office - Chair water committee - Woreda agricultural office, irrigation department - Woreda agricultural office, cooperative department	4
B: Raya-Azebo diversion	- Woreda agricultural bureau - Woreda water bureau - 2 farmers interviews Tsigea kebelle - Committee members + farmers from Genete kebelle	5
C: Kugitelemlem vs. Selamgelasi	- Woreda agricultural bureau - Woreda water bureau - 2 times interview with committee members and farmers from the irrigation system	4
D: Berki watershed	- Woreda water bureau Atsbi-Womberta - Woreda agricultural bureau Atsbi-Womberta - Woreda water bureau Kilite-Awlalo - Woreda agricultural bureau Kilite-Awlalo - Three times farmer interview in Atisbi-Womberta woreda (one diversion, one hand dug well and one time river irrigation with a motor pump) - Two farmer interviews in Kilite Awlalo woreda (river irrigation with a motor pump and diversion) - One interview with a village chairperson from a kebelle in Kilite-Awlalo woreda.	10
E: Degem	- Woreda water bureau - Woreda agricultural bureau - Four members of the water committee in Tumano - Seven different farmer interviews in Tumano kebelle - Five different farmer interviews in Ano Kare kebelle - One interview with a member of the Ano Kare water committee	16
F: Diversion Kobo	- Woreda agricultural office, irrigation department - Woreda agricultural office, cooperative department - KGVDP office - Former committee member and now farmer of the diversion in Golina kebelle	4
G: Diversion Kombolcha	- Woreda agricultural bureau - Water Action local office - Eight members of the water committee	3

Appendix III Interviews used in the thesis

In this appendix, the interviews referred to in the thesis are listed.

Table a: list of interviews referred to in the thesis

Interview number	Function interviewee(s)	Place and date of interview
Interview 1	Agronomy expert and extension coordinator from the Alamata Woreda Agricultural office	Alamata, 1-03-2013
Interview 2	Kebelle chairperson of Tesoma kebelle about spate irrigation schemes	Tesoma kebelle, Alamata, 3-03-2013
Interview 3	Irrigation expert of the Kobo-Girana Valley Development Programme	Kobo, 4-03-2013
Interview 4	Chair of the water committee in Aybe, modern irrigation scheme	Aybe kebelle, Kobo, 4-03-2013
Interview 5	Three farmers using hand dug wells	Atsbi-Womberta, 8-03-2013
Interview 6	Irrigation official of the Ministry of Water and Energy	Addis Ababa, 27-03-2013
Interview 7	Director of the Irrigation and Drainage department of the Oromia Regional Water Bureau.	Addis Ababa, 28-03-2013
Interview 8	Irrigation Engineer of the Amhara Regional Water Bureau	Bahir Dar, 15-04-2013
Interview 9	Director of irrigation department of the Amhara Regional Agricultural Bureau	Bahir Dar, 15-04-2013
Interview 10	Eight members of the water committee in 09 kebelle	Kombolcha, 9-04-2013
Interview 11	Director of Irrigation Department of the Ministry of Water and Energy	Addis Ababa, 8-05-2013
Interview 12	Agronomist of the Raya-Azebo woreda agricultural bureau	Mohoni, Raya-Azebo, 6-03-2013
Interview 13	One large-scale farmer from Tsigea kebelle	Tsigea kebelle, Raya-Azebo, 6-03-2013
Interview 14	Water committee members and farmers from Genete kebelle (eight men)	Mohoni, Raya-Azebo, 6-03-2013
Interview 15	Director of the Alamata woreda water bureau	Alamata, 1-03-2013
Interview 16	Water committee members and farmers of Fascha water point (six men)	Kugitelemlem kebelle, Alamata, 2-03-2013
Interview 17	Extension expert of the Kilite-Awlalo woreda agricultural bureau	Wukro, Kilite-Awlalo, 7-03-2013
Interview 18	Head of the Atsbi-Womberta woreda agricultural bureau	Atsbi-Womberta, 8-03-2013
Interview 19	Agronomist of the Atsbi-Womberta woreda water bureau	Atsbi-Womberta, 8-03-2013
Interview 20	Head of drinking water and irrigation process of the Kilite-Awlalo woreda water bureau	Wukro, Kilite-Awlalo, 9-03-2013
Interview 21	Three farmers using irrigation from the Berki river	Kilite-Awlalo, 9-03-2013

Interview 22	Interviews villagers in Tumano kebelle (eight men and two women)	Tumano kebelle, Degem, 1-04-2013
Interview 23	Two irrigation experts from the Degem woreda water bureau	Degem, 1-04-2013
Interview 24	Farmers from Ano Kare kebelle using water from the spring originating in Tumano (three men)	Ano Kare kebelle, Degem, 3-04-2013
Interview 25	Former chair of the water committee managing the diversion in Golina kebelle, Kobo	Golina kebelle, Kobo, 4-03-2013
Interview 26	Eight members of the water committee in 09 kebelle in Kombolcha	09 kebelle, Kombolcha, 9-04-2013
Interview 27	Irrigation expert of the Kombolcha woreda agricultural bureau	Kombolcha, 9-04-2013
Interview 28	Water action project office Kombolcha	Kombolcha, 9-04-2013
Interview 29	Irrigation expert of Kobo woreda agricultural bureau	Kobo, 5-03-2013
Interview 30	Three farmers from Tsigea Kebelle	Tsigea kebelle, Raya-Azebo, 6-03-2013
Interview 31	Water committee member of Biruhtesfa irrigation scheme in Alamata woreda	Selamgelasi kebelles, Alamata woreda, 2-03-2013
Interview 32	Four members of the water committee and three farmers from a furrow irrigation scheme in Jarota kebelle, Kobo	04 Jarota kebelle, Kobo woreda, 4-03-2013
Interview 33	Four water committee members and six farmers of a modern system in Alamata woreda	Alamata woreda, 2-03-2013
Interview 34	Two operators from a modern system in Golina kebelle, Kobo	08 Golina kebelle, Kobo woreda, 4-03-2013.
Interview 35	Two farmers using a modern system in Worgeba kebelle, Raya-Azebo	Worgeba kebelle, Raya-Azebo woreda, 6-03-2013
Interview 36	Village Chairperson of Abreha wa Atsbha kebelle	Wukro, Kilite-Awlalo woreda, 10-03-2013
Interview 37	A local development agent and two farmers of a diversion in Alewoha kebelle, Kobo	Alewoha kebelle, Kobo woreda, 5-03-2013
Interview 38	Data official of the Regional Water Bureau	Mekelle, Tigray 14-03-2013
Interview 39	Irrigation and water resource management expert of the NGO Water Action	Addis Ababa, 27-03-2013
Interview 40	Head of the irrigation department of the Degem woreda water bureau	Degem woreda, 3-04-2013
Interview 41	Professor in hydrology of the Mekelle University	Mekelle, Tigray, 7-03-2013